Real Estate Math

Completely Explained

Gerald Shingleton

Real Estate Math

Completely Explained

CadArm Publications
Casa Grande, Arizona 85122
Copyright © 2012   Gerald Shingleton
All rights reserved.
ISBN-10: 1478315814
ISBN-13: 978-1478315810

BOOKS BY
GERALD SHINGLETON

- **JOURNEY TO TERRAINCOGNITA**

- **JUNEBUG**

- **GOD POWER**

- **ARCHITECTS REFERENCE MANUAL**

- **RAINBOW CAPER**

Chapter Content

AUTHOR'S INTRODUCTION

I wrote this book because of my experience in Real Estate School. Students in pursuit of a career as a licensed agent must pass a state sponsored exam. One particular topic, it would seem, was tough, yet required.

Lack of understanding and in particular, the importance of units, on the part of the instructor, who in fact owned the school, was frustrating. She insisted the one topic that required her expertise was real estate math. This was particularly alarming when lineal feet and square feet got mixed up in her mind with little understanding of how they differ when computing area.

Students smiled, or stared glass-eyed, and agreed with her assessments and solutions to a particular problem. The answer was completely wrong, misleading due to ignorance. I gave up explaining that units were being misunderstood, especially when no other student in a classroom of thirty supported my contentions. It was frustrating being viewed as a know-it-all, because who doesn't trust an experienced teacher in the business of assisting future agents?

But all I witnessed were frustrated students with a complete lack of fundamental understanding. Most actually took that class more than once and still failed to grasp the material. Text books were way too general and assumed the reader had a basic understanding of fractions and units. Sadly, that's not the case. So, I thought to myself, perhaps a very clear and simplified description and explanation of the subject was in order.

I also felt a responsibility to help agents understand practical mathematical language and be versed on a most important aspect of the profession they are a part of. Now, you may find it strange that there are plenty of math made simple books, so why add another to the mix? Most are written by industry leaders who assume the reader is brushing up on a subject once thoroughly known. The problem still exists for the frustrated student who never really grasped math well while growing up.

The way this book is structured is completely different. Each subject is discussed in a chronology to make everything come together. Start from the beginning and make sure each particular subject is understood before going to the next. Finally, test questions have complete explanations about how to solve the given problem.

1 UNITS

What is a unit? What does that have to do with arithmetic? You'll learn this term has everything to do with understanding real estate mathematics.

Here's an example of how important it is. Imagine you have ten apples. The word *apple* becomes a unit; a referenced attachment to a number. You also have ten fingers; the word finger is a unit attached to a number. If you have two apples and four pears, the total equals 6 pieces of fruit. In this case, the unit attachments must absolutely change to suit the explanation.

Everyone has heard mention of feet and inches. Those are units of measurement. Dollars and cents are financial units. When you weigh yourself, pounds and ounces are units as well. Everything talked about relating to size, weather, cooking, quantities, time, projections, money, and financing, have units.

When you go to a doctor and step on a scale,

your weight is recorded. If you do not know what units are used, it would be very difficult to understand the answer. Like any language, there has to be an understanding or communication is impossible. Is it kilograms, ounces, pounds, English pounds? If the doctor says you weigh 125. That's usually understood to mean 125 pounds. The unit is further confused by nomenclature because there are different ways of expressing this; like 125#, or 125 lbs.

But, what if you were in a different country and the doctor claimed you weighed 57. A translation is mandatory so there's an understanding. If he didn't mention units, would you assume that you lost weight or maybe challenge the accuracy of the device? If you discover the weight is 57 kilograms, would that make sense? Probably, just like a communication barrier or mistranslation, the answer would seem useless unless you understood the difference between the measurement units.

Understanding units in real estate language is extremely important. As an example; a property frontage is claimed to be exactly 40 meters in length. What does that mean to you? Have you heard of meters? Or do you picture in your mind feet or yards? There again is a failure to communicate, exactly like a language barrier, especially if you can't translate that 40 meters is about 110 feet.

By the way; mentioning the metric system deserves further explanation. This particular mathematical language makes complete sense and is by far the easiest to learn and understand. That's because all units are based on the root ten and simplified math also uses the root ten. So, the question remains, "why isn't the metric system used universally? Furthermore, what countries besides the U.S. have not adopted the metric system?"

Many U.S. teachers think the answer is "Liberia and Burma" but that's no longer true. All countries have adopted the metric system, including the U.S., and every country, except the good old United States; have taken steps to eliminate most uses of traditional unit measurements. However, in nearly all countries people still use traditional units sometimes, at least in colloquial expressions.

Becoming metric cannot become a one-time event that either happens or not. It has to be a process that happens over time. Every country is somewhere in this conversion, some much further along than others.

To give you an idea of just how complex this is, imagine if the official language were changed; mandated by Congress to be Spanish in lieu of English. Ordering products and fixtures would require retooling and everyone's new thought processing could get mixed up. For example; a bathtub would be 1.5 meters in length instead of 5 feet.

I'm sure by now, units mean everything and without accepting their utmost importance, real estate math becomes meaningless. So, just for the fun of it, let's investigate the very important and most common ones for a greater appreciation. Speaking in the language most understood in the United States, the imperial system of units is used. Thus, the world recognized and most logical metric system is set aside, for now, anyway.

FOOT:

Can you believe this basic unit of measurement wasn't even officially defined internationally until 1959? Former definitions were confusing from country to country. The problem became significant when recording large surveys because distances actually meant something different, especially between the United States and India, for example.

Historically the human body was used to provide the basis of this measurement unit. Scientifically, the foot of a Caucasian male was determined to average about 15.3% of his height. Forensic examinations could measure a footprint and come up with an approximate height for a suspect.

Shoes replaced primitive sole-wear and were designed and originally called a foot and a pair of shoes would be feet. The actual foot length became standardized and evolved into measuring. Before the advent of shoes, the ancients used cubits, but it was the Romans and Greeks who developed a different means of measuring after the advent of footwear.

Just for fun, measure a size 10 or 11 man's shoe and see how close to one foot it is. You may be surprised at the results and in some cases it could be so close to being exactly one foot, you could rely on the accuracy in measuring the dimensions of a room when showing a home.

At first, the foot measurement was divided into 16 separate sections, but later decided that 12 sections (or unicas) correlated better, especially since there were 12 months to a year. Oh well, now we know something that others take for granted. But, the foot measurement is probably the most significant of all units in real estate.

YARDS:

The origin of the Yard isn't really known. Perhaps it derived from the double cubit, or from its near equivalents, such as the length of a stride or pace. Another possibility was the yard was derived from the girth of a person's waist. Another claim held that the measure was invented by Henry I of England as being the distance between the tip of his nose and the end of his thumb. The unit is a little confusing because of technical misuse. The word is commonly used as a homonym in the sense of an enclosed area of land. There's also a confusing use where the unit means volume; like 3 yards of concrete. It could be a specific area as well; like 4 yards of fabric.

So there are corresponding units of area and volume: the square yard and cubic yard. These are commonly referred to simply as yards when no ambiguity is possible. For example, an American or Canadian concrete mixer may be marked with a capacity of 10 yards, where cubic yards are obviously referred to.

Now this can be real confusing since units expressed as slang can be misleading. As you will later learn, correct terminology would be 4 cubic yards of concrete or 2 square yards of cloth. The quarter of a yard used to be known as the quarter without any qualification, while the sixteenth of a yard was called a nail. The eighth of a yard was sometimes called a finger, but was more commonly referred to simply as an eighth of a yard, while the half-yard was called half a yard.

Other units related to the yard, but not specific to cloth measurement: two yards are a fathom, a quarter of a yard (when not referring to cloth) is a span.

The yard is used as the standard unit of field-length measurement in American, Canadian and Association football, cricket pitch dimensions, swimming pools, and in some countries, golf fairway measurements.

INCH:

The Latin word *unica* means one twelfth part, just like the word ounce originally represented one twelfth of a troy pound. The English word inch was derived from this same idea. In many other languages, the word for inch is similar or the same as the word thumb.

King David I of Scotland in his Assize of Weights and Measures (c. 1150) defined the Scottish inch as the width of an average man's thumb at the base of the nail. An Anglo-Saxon unit of length was the barleycorn. After 1066, 1 inch was equal to 3 barleycorns, which continued to be its legal definition for several centuries, with the barleycorn being the base unit. One of the earliest such definitions is that of 1324, where the legal definition of the inch was set out in a statute of Edward II of England, defining it as "three grains of barley, dry and round, placed end to end, lengthwise".

ACRE:

Real estate math typically uses the word acre. When you attend a live American Football game, the playing field is smaller than one acre, but not a whole lot. So, in essence, the word describes an area. So, for reference purposes think of an acre as a unit of area.

The most commonly used acres today are the international acre and, in the United States, the survey acre. Typically, the acre is used to measure tracts of land. During the Middle Ages, an acre was the amount of land that could be plowed in one day with an ox.

Real estate professionals use other terms related to the word acre. For example, the term acre-feet can explain a volume, usually a body of water. So imagine a football field one foot deep filled with water. As a rule of thumb in U.S. water management, one acre-foot is taken to be the planned water usage of a suburban family household, annually. In some areas of the desert southwest, where water conservation is followed and often enforced, a typical family uses only about one quarter acre-feet of water per year.

The acre is often used to express areas of land in the United States and in countries where the Imperial System is still in use. As of 2010, the acre is not used officially in the United Kingdom. In the metric system, the hectare is commonly used for the same purpose. An acre is approximately 40% of a hectare.

SECTION:

In U.S. land surveying under the Public Land Survey System (PLSS), defines a section as an area nominally one square mile, containing 640 acres, with 36 sections making up one survey township on a rectangular grid. Legal descriptions of a tract of land under the PLSS includes the name of the state, name of the county, township number, range number, section number, and portion of a section. Sections are customarily surveyed into smaller squares by repeated halving and quartering. A quarter section becomes 160 acres and a "quarter-quarter section" is 40 acres. In 1832 the smallest area of land that could be acquired was reduced to the 40 acre quarter-quarter section, and this size parcel became entrenched in American tradition. After the Civil War freed slaves were reckoned to be self-sufficient with "40 acres and a mule." In the 20th century, real estate developers preferred working with 40 acre parcels. The phrases "front 40" and "back 40," referring to farm fields, indicate quarter-quarter sections of land.

One of the reasons for creating sections of 640 acres was the ease of dividing into halves and quarters while still maintaining a whole number of acres. A section can be halved seven times in this way, down to a 5 acres parcel, or half of a quarter-quarter-quarter section—an easily surveyed 50 square chain area. This system was of great practical value on the American frontier, where surveyors often had a shaky grasp of mathematics and were required to work quickly.

The existence of section lines made property descriptions far more straightforward than the old metes and bounds system. The establishment of standard east-west and north-south lines ("township" and "range lines") meant that deeds could be written without regard to temporary terrain features such as trees, piles of rocks, fences, and the like, and be worded in the style such as "Lying and being in Township 4 North; Range 7 West; and being the northwest quadrant of the southwest quadrant of said section," an exact description in this case of 40 acres, as there are 640 acres in a square mile.

The importance of "sections" was greatly enhanced by the passage of "An Ordinance for ascertaining the mode of disposing of lands in the Western Territory" of 1785 by the U.S. Congress. This law provided that lands outside the then-existing states could not be sold, otherwise distributed, or opened for settlement prior to being surveyed. The standard way of doing this was to divide the land into sections. An area six sections by six sections would define a township. Within this area, one section was designated as school land. As the entire parcel would not be necessary for the school and its grounds, the balance of it was to be sold, with the monies to go into the construction and upkeep of the school.

DAY:

A day is a unit of time, commonly defined as an interval equal to 24 hours. The period of time measured from local noon to the following local noon is called a solar day.

Several definitions of this universal human concept are used according to context, need and convenience. The word day may also refer to a day of the week or to a calendar date, as in answer to the question "On which day?" Day also refers to the part of the day that is not night — also known as 'daytime'. The life patterns of humans and many other species are related to Earth's solar day and the cycle of day and night.

It is interesting to note that the Earth's day has increased in length over time. This phenomenon is due to tides raised by the Moon which slow Earth's rotation; increasing by about 1.7 milliseconds per century.

Real estate math ignores precision and simply assumes the DAY unit is precisely one thirtieth of a month and there are exactly 360 day units comprising one full year. Professionals refer to this as a statutory year. How crazy is that? Shortcuts like this really don't exist in the financial institutions where long term precision is mandatory.

***

Units are as important as numbers. One way to look at it is very simple; some figures are just numbers (quantitative) and some are numbers with units (numerical attachments). For example, one dollar is the number one with a unit attachment. If your friend has twice as much, the total would be two dollars. This assumption basically says; $1 x 2 = $2. Thus there's one figure with a unit attachment and one figure without any unit.

A *rule-of-thumb* is important to explain here. Plant this solidly in your mind; this will help you when faced with a mathematical problem or a test question. Units of MEASURE behave differently than others. Lineal measurements can be multiplied together. The result becomes an area measurement. Area measurements can be multiplied by a lineal measurement. The result becomes a volume measurement. If units are NOT measurements, they cannot be multiplied together.

In this example, it is impossible to multiply $1 x $2. Why? The answer is rather simple. When units are multiplied, the answer must reflect a square area. But, there's no such thing as square dollars (obviously). 1FT x 2FT = 2 square feet is a proper answer because linear feet x linear feet is an area measurement.

Multiplying the same units or dividing the same units is possible only when discussing real estate math. Keep this rule of thumb in mind and always ask yourself; "If I multiply these units together, will the result be an area?" If the answer is no; obviously it cannot be done. Remember; adding like units together is absolutely possible; like 4 ft plus 4 ft equals 8 ft. Subtracting like units is practical; 8 ft minus 4 ft equals 4 ft.

Never mix units together without converting. For example; 3 ft times 3 yards can be done only if the units match. This can be done simply by converting both figures to feet or yards, whichever is convenient; 1 yard times 3 yards equals 3 square yards. The same answer is achieved by multiplying

3 ft times 9 ft equals 27 square feet. Both answers are the same. 27 square feet is the equivalent of 3 square yards.

There's another *rule-of-thumb* to consider as well. ALL identical units can be divided but the answer will never have a unit attached. That's because any unit divided by the same unit will cancel out. So, for example; 2 feet divided by 1 ft is equal to 2.

Division is the opposite of multiplying so reversing the process double checks an answer. In this case, 1 ft times 2 is equal to 2 feet.

The next chapter will discuss numbers with no units which, at least in real estate math, are generally expressed as a ratio or percentage. What is crucially important to understand, is being like minded. In other words, always make sure units are like minded (match).

If a problem or test question seeks a ratio or percentage, the like minded units somehow must cancel out. That can only be done by dividing. Adding, subtracting, or multiplying will NEVER cancel units out! Just knowing and understanding this principle will no doubt assist in determining a correct answer.

Acre = 43,560 square feet (SF) (area measurement)

Mile = 5,280 feet (linear measurement)

Square Mile = 5,280 ft. x 5,280 ft. = 27,878,400 sq. ft. = 27,878,400 sq. ft. / 43,560 sq. ft. per acre = 640 acres.

Yard = 3 feet (linear measurement – except when used as slang)

Square Yard = 3 ft x 3 ft = 9 square feet (area measurement)

Cubic yard = 3 ft x 3 ft x 3 ft = 27 cubic feet (volume)

Section = 640 acres = 1 square mile (area measurement)

Square Foot = 12 inches x 12 inches = 144 square inches (area measurement)

The Metric System

There's a possibility that a test question will require conversion to the world standard metric system. Code books and official references are currently including the metric system as a way to eventually transfer the American mathematics language to a worldly understanding.

The system makes complete sense, as previously discussed, because the root "ten" is the basis for common mathematics. Ten fingers and ten toes establish the basic principle, like counting by fives or tens. Here presented are the basic terms.

Kilo   means 1,000

hecto   means 100

deca   means 10

deci   means 0.1

centi   means 0.01

milli   means 0.001

Conversion Table

Kilometer	km	1,000 meters	0.62 miles
Hectometer	hm	100 meters	328.08 ft.
Decameter	dam	10 meters	32.81 ft.
Meter	m	1 meter	39.37 inches
Decimeter	dm	0.1 meter	3.94 inches
Centimeter	cm	0.01 meter	0.394 inches
Millimeter	mm	0.001 meter	0.039 inches

1 yard = 0.9144 meter = 0.5 fathoms = 3 ft.

2 PERCENT

What is a percent? What does that have to do with real estate? You'll learn that this term has everything to do with understanding real estate financing and area ratios. The word is represented by a symbol (%).

Imagine, if you will, a couple rolls of pennies all lying on a table top. Each coin is the equivalent of one cent. It is easy to ascertain that each roll has fifty pennies, so two rolls total 100 cents. That's the same as one dollar. Think about that; one dollar is the same as 100 cents.

Percent is similar to cents. In a way, it is just like a unit of measure, the same as a penny. The same principles apply, well not exactly. The truth is; a percent is really a number without a unit once converted properly. So just like 100 cents is one whole dollar, 100% is one whole number.

Most people understand how money works. Everyone knows that a nickel is worth less than a dime even though a nickel is larger. A quarter is worth less than a half dollar. All of these can be converted to numbers with units. For instance; if you have a one dollar bill, four pennies, and two quarters, the amount can be simplified by converting to one unit. Normally, the unit used and understood is the dollar ($).

Results are basic; $1.54 (one point five four dollars or a dollar fifty four or one dollar and fifty four cents). Percent, like cents, works the same way. Let's say you have a dollar and someone offers to increase the value by four percent. Then two others claim they will add another twenty five percent. The value adds up to fifty four percent. What you end up with is 1.54% more than you started with.

Percent can be imagined that way, as cents expressed strictly as numbers without units. One percent is the number 0.01 (just like one cent is $0.01 – a number with a unit, in this case).

The word is derived from the Latin per centum meaning "by the hundred". For example, 15% (read as "fifteen percent") is equal to 15 divided by 100 or whole or expressed as 15/100, or 0.15.

A lot of calculators have a percent key (%). Most do not. It is not recommended unless you really understand how it works. It is better suited to convert a percentage into a number before doing a computation. That's because percentage is just another way of expressing a number (no units attached), especially a ratio, as a fraction of 100.

Percentages are used to communicate how large/small one quantity is, relative to another quantity. The first quantity usually represents a part of, or a change in, the second quantity, which should be greater than zero. For example, an increase of $0.15 on a price of $ 2.50 is an increase by a fraction of $0.15/2.50 = 0.06$. Expressed as a percentage, this is therefore a 6% increase. Notice that the decimal point is moved two places to the right to convert a plain number to the percent unit.

Although percentages are usually used to express numbers between zero and one, any ratio can be expressed as a percentage. For instance, 111% is 1.11 and 0.35% is 0.0035. So you see, conversion from a number expressed as % unit means moving the decimal point two spaces to the left when converting percent to a number. That particular operation basically means dividing the decimal number by 100. Likewise, converting a number to a percent requires moving a decimal point two spaces to the right. That operation basically means multiplying the decimal number by 100. Remember, a decimal point, if not depicted is assumed to be at the end of a number.

All industry agents have hand held calculators. The device is typically allowed when taking state exams. Practice converting percents by utilizing the magic "whole" number of 100. Coincidentally, when working with dollars and cents, converting pennies into dollars utilizes the same principle, because 100 pennies makes one "whole" dollar.

In the case of interest rates, it is a common practice to state the percent change differently. If an interest rate rises from 10% to 15%, for example, it is typical to say, "The interest rate increased by 5%" — rather than by 50%, which would be correct when measured as a percentage of the initial rate (i.e., from 0.10 to 0.15 is an increase of 50%).

You can probably imagine just how confusing that would be in the industry. Even though mathematically correct, discussing two different percentages (units) all mixed together could cause misunderstanding. Such ambiguity can be avoided by using the term "percentage points". In the previous example, the interest rate "increased by 5 percentage points" from 10% to 15%. If the rate then drops by 5 percentage points, it will return to the initial rate of 10%, as expected.

The actual percent value, further emphasized, is computed by multiplying the numeric value of the ratio by 100. Basically, as discussed prior, that's what happens when a decimal point is moved two spaces. For example, to find the percentage of 50 apples out of 1250 apples, first compute the ratio 50/1250 = .04, and then multiply by 100 to obtain 4%. The percent value can also be found by multiplying first, so in this example the 50 would be multiplied by 100 to give 5000, and this result would be divided by 1250 to give 4%.

If you are still confused, think again about how pennies are cents which are like percents. One hundred cents is a "whole" dollar which is 100¢ expressed as $1.00.

To calculate a percentage of a percentage, convert both percentages to fractions of 100, or to decimals, and multiply them. For example, 50% of 40% is:

(50/100) × (40/100) = 0.50 × 0.40 = 0.20 = 20/100 = 20%.

It is not correct to divide by 100 and use the percent sign at the same time. (E.g. 25% = 25/100 = 0.25, not 25% / 100, which actually is (25/100) / 100 = 0.0025.) That's why a pocket calculator with a percent key is not recommended, it can get too confusing.

There's an easy way to calculate additions in percentages (discount 10% + 5%): For example, if a department store has a "10% + 5% discount," the total discount is not 15% because the actual total discount must be relative. That's because, if a jacket is priced at $100 with an advertised discount of 10% and then the article is again offered with a 5% clearance sale, the final price is not 15% less. Here's how to figure the selling price.

Convert 10% = 0.10 (always convert % to a number)

$100.00 x 0.10 = $10.00 discount

Base price = $100.00 - $10.00 = $90.00

Convert 5% = 0.05

$90.00 x 0.05 = $4.50

Sales Price = $90.00 -$4.50 = $85.50

FYI: There's another way to determine the final sales price, expressed as an algebraic calculation.

$\{1.0 - [(0.10 + 0.05) - (0.10 \times 0.05)]\} \times \$100.00 = \$85.50$

Whenever we talk about a percentage, it is important to specify what it is relative to, i.e. what is the total that corresponds to 100%. The following problem illustrates this point.

In a certain college 60% of all students are female, and 10% of all students are computer science majors. If 5% of female students are computer science majors, what percentage of computer science majors are female?

I wouldn't throw my arms in the air and give up. You could pretend the percent signature was gone, in this case. Remember a whole dollar is 100 cents. So, think of the whole population as being 100 students. That would mean there would be 60 females. There are also 10 computer science majors. Five percent (5%) converts to a number which equals 0.05. 60 females times 0.05 = 3 females who are indeed computer science majors. Now if there are 100 students total and 3 female computer science majors, the ratio is 3 per 100 which is expressed as 3 divided by 100 = 0.03 or 3% of all students are female computer science majors. But really, the answer to the question is 3 female students divided by 10 total computer science majors is equal to 0.30 or 30%.

There's another way to figure this out, of course. We are asked to compute the ratio of female computer science majors to all computer science majors. We know that 60% of all students are female, and among these 5% are computer science majors, so we conclude that (60/100) × (5/100) = 3/100 or 0.60 x 0.05 = 0.03 = 3% of all students are female computer science majors. Dividing this by the 10% of all students that are computer science majors, we arrive at the answer: 3%/10% = 0.03 / 0.10 = 0.30 or 30/100 or 30% of all computer science majors are female.

This example is closely related to the concept of conditional probability. Sometimes due to inconsistent usage, it is not always clear from the context what a percentage is relative to. So that's why separating out the different types can be visually supportive.

When speaking of a "10% rise" or a "10% fall" in a quantity, the usual interpretation is that this is relative to the initial value of that quantity. For example, if an item is initially priced at $200 and the price rises 10% (an increase of $20), the new price will be $220. Note that this final price is 110% of the initial price (100% + 10% = 110%). In essence, 110% = 1.10 and multiply this number by the original price of $200 equates to $220.

Some other examples of percent changes:

An increase of 100% in a quantity means that the final amount is 200% of the initial amount (100% of initial + 100% of increase = 200% of initial); in other words, the quantity has doubled.

An increase of 800% means the final amount is 9 times the original (100% + 800% = 900% = 9 times the original amount).

A decrease of 60% means the final amount is 40% of the original (100% − 60% = 40%).

A decrease of 100% means the final amount is zero (100% − 100% = 0%).

It's crucial and important to understand that percent changes, as they have been discussed here, do not add in the usual way, if applied sequentially. For example, if the 10% increase in price considered earlier (on the $200 item, raising its price to $220) is followed by a 10% decrease in the price (a decrease of $22), the final price will be $198, not the original price of $200. The reason for the apparent discrepancy is that the two percent changes (+10% and –10%) are measured relative to different quantities ($200 and $220, respectively), and thus do not "cancel out".

Another common mistake is thinking that working 50% faster means taking 50% less time to complete the task. On this account, 100% faster means twice the speed, so half the time. For example, if one traveled at 50 mph, 100% faster would be 100 mph (taking 50% less time). And 50% faster speed means 33.33% less time to travel the same distance; because 0.50 + 1.00 = 1.50 or 150%. 50 mph x 1.50 = 75 mph. The ratio 50 mph divided by 75 mph equates to 66.66% which is subtracted from the whole (or 100%); making it 1/3 or 33.33% less time to travel. Another way to decipher and still get the same results is by subtracting the original 50 mph from the resulting 75 mph to get 25 mph. 25 mph divided by 75 mph gives the same final result of 33.33%.

3 MAGIC FORMULA

Most students of real estate have no training in algebra or perhaps nothing beyond basic math. So, the way classes are taught, the teacher diagrams a magic formula circle. But before even discussing this aspect, there are a few more *rules of thumb* to consider and firmly plant in the mind.

A fraction is a number just like any, but it is written with one number over another. An example is ½ or 1 over 2 which simply means half of a whole. There's another way this fraction can be written; ½ is the same as 0.50. The way a fraction can be converted is to take the top number and divide it by the bottom one (numerator divided by the denominator). Using a calculator, enter 1 and then hit the division button (÷), and then enter the number 2. The calculator's answer is 0.5.

Just to make this clear, the fraction ¾ can be converted to a standard number by taking 3 and dividing it by 4 which equals 0.75. In essence, fractions are probably a lot easier to add or subtract by first converting them into whole decimal numbers. For example; add ¾ plus ½. These fractions cannot be added until the denominators (lower number) matches. The lowest common denominator is the number 4, which means that ½ is changed to 2/4 before it can be added to ¾. So the answer is 5/4 which is greater than 1. One more step is necessary; subtracting 4/4 as a whole number, making the final answer 1 ¼.

An easier way is to convert to decimal; ¾ = 0.75 and ½ = 0.5. Thus; 0.75 + 0.5 = 1.25 (the same as 1 ¼).

Keep in mind that whenever you have a number over another number, it basically means divide. In real estate math, generally speaking, the top number will be smaller than the lower number. So if it is logical that the final result is clearly a fraction or less than the whole, this will certainly be the case.

GENERAL RULES OF THUMB

1. A part of something divided by a percent = a whole
   Example: One dollar is a whole amount. If all you have left after spending 25 cents would be 75 cents. The percentage can be figured out because the part of something is 75 cents and the whole is $1.00.
   $0.75 ÷ \% = 1.00$; Thus $\% = 0.75 = 75\%$ (amount remaining)

2. A part of something divided by the whole is a part or percent.
   Example: One dollar is again the whole amount. If 25 cents is spent, the percentage spent is expressed simply by taking the part spent and dividing it by the whole.
   $\$0.25 ÷ \$1.00 = 0.25 = 25\%$

3. The whole times the percent is equal to the part.
   Example: One dollar is a whole and the percentage spent is 25 %; thus $\$1.00 \times 25\% = \$1.00 \times 0.25 = \$0.25$

4. When a unit is divided by a unit, the answer is always a number without a unit.
   Example: 1 ft ÷ 2 ft = 0.5 or 50%. Another

sample; $25.40 ÷ $4.25 = 5.98 = 598%

5. When a unit of like kind is divided by a like kind, convert and cancel out to get the correct unit in answer.
Example: One Acre ÷ 5,280 ft = 43,560 sq. ft. ÷ 5,280 ft =43,560 ft x ft ÷ 5,280 ft = 8.25 ft (remember ft x ft is sq. ft. and ft ÷ ft = 1)

6. The words "less, loss, after, discount, deduct," are specific clues in a word problem which *almost* always means to subtract a given percentage from 100%.

P A R T
(WHOLE) | (%)

The schematic above is a key diagram typically used to teach students how to determine a method for correctly answering a word problem. If the question asks to find the part or the smaller portion of a given problem, simply cover the word PART with your thumb. The solution is to **multiply** the whole by the percentage (remember to convert to a number).

If the part or smaller portion of the given

material is available, but the percentage is unknown, cover the (%) and see that the answer is determined by **dividing** the smaller part by the whole. Remember, the answer will be a decimal number where the decimal point must be moved two places to the right for percent.

If the percentage is given, convert to a decimal by moving the decimal point to the left two integers. Then, cover up the part or the whole, depending on which is given in the problem. If the whole is unknown, then the answer is part **divided** by the percent (%).

Here are a few test questions to make sure you understand the method.

***

The house sells for $100,000 and the sales agent receives $1500 commission. What's the commission percentage?

Answer method:

Step 1: There are 3 factors indicated so the ingredients match the schematic. Thus no further breakdown or elimination is relevant.

Step 2: Compare units for no conflict; $ (dollars) are given for both amounts. Thus no conversion is necessary.

Step 3: Obviously, the answer will be an amount less than one whole (1.00). Thus, the lower amount is the part which belongs at the top of the schematic, the higher number at the bottom.

Step 4: Cover the (%) symbol; thus the answer is the top figure divided by the bottom one.

Step 5: $1500 ÷ $100,000 = 0.015 (remember the $÷$ = 1 or cancels out)

Step 6: Convert answer to percent by moving the decimal two places to the right. 0.015 = 1.5% (final answer)

***

A seller wants to net exactly $125,000 from the sale of a home to repay the existing encumbrance on the property. He must pay a 6% broker's commission. What will the home have to sell for?

Answer method:

Step **1**:

There are 3 factors (2 of the 3 or given) indicated so the ingredients match the schematic. Thus no further breakdown or elimination is relevant; except the problem asks for a gross sale amount when the net sale amount is given. This should raise a red flag and perhaps another *rule of thumb* is relevant. In other words, the commission is based on the gross amount and not the net, or what's left over after the commission is "subtracted".

Keep in mind that whenever there's "subtraction" necessary, setting up the correct figures will require some subtraction.

Step **2**:
So in this case, the given percentage is tricky because what is necessary to obtain the answer is the seller's **net** percentage, not what the broker makes on the deal. Even though clue words (#6 of general rules of thumb) are not specific, there are insinuations about being less. The net sale is "less" than the gross sale amount. Caution; never assume or carelessly ignore what exactly is being asked. A common error is to assume that all information is final without further assessment. The *real* percentage to solve the problem requires a subtraction step which simply is 100% - 6% = 94% = 0.94. (That number goes in the lower right of the magic formula.)

Step **3**:
The $125,000 amount is **part** of the whole, so it belongs at the top (generally a rule to follow).

Step **4**:
The unknown is the **whole** (in dollars), which Is a question mark or unknown at lower left of the magic formula. So cover it with a thumb and see the answer: $125,000 ÷ 97%

Step 5:
Convert percentage to a whole number
$125,000 ÷ 0.94 = $132,978.72. (final answer)

Step 6:
Never be satisfied with coming up with an answer. Always check the results by reversing the procedure. It's so easy to do, and takes little time to work everything back to make sure figures are correct. $132,978.72 is the assumed gross sale amount. So, if correct, subtracting a 6% sales commission should result in the net amount required of $125,000.
$132,978.72 X 6% = $132,978.72 X 0.06 = $7,978.72.
So the sales agent receives this amount but the seller gets what's left over.
$132,978.72 - $7,978.72 = $125,000 (okay).

Checking work is best for even the experienced real estate math wizard. One never knows when a slight error may occur, like inadvertently entering a value of 97% instead of 94%. That could happen to anyone so working backwards will reveal errors.

4 MATH PROBLEM BASICS

This chapter is a basic outline and further review of necessary steps and procedures used to solve a basic math problem that would most likely be encountered on a state examination; Real Estate mathematics standard rules and steps. Learn these as a matter of habit.

Word problems contain units (measurement-time-percentage-weight-etc.). Never assume all units "mesh". For example, yards and feet do not work at all together. Yards have to be converted into lineal feet (do not make the mistake of converting lineal yards into square feet! That's a completely wrong assumption and misunderstanding of units). Pounds and ounces also do not mesh together, so convert pounds into ounces. Years, weeks, and days can also be converted to mesh as days so all units are identical.

Learn (memorize) the following rules, conversions, principles, and unit measurements; crucial principles for solving basic real estate math **measurement** problems.

1 Yards can be converted into feet (lineal feet) by multiplying by 3.

One yard is 3 feet (think of a yardstick). Mathematically: 1 yd. / ft. x 3 ft. / yd = 3

FYI: Remember, the word *yards* used in the building industry is slang. For example; 7 yards of concrete is actually cubic yards. Two yards of gravel is 2 cubic yards.

Carpeting is measured in yards as well. But that too is slang. Five yards of carpeting is actually five square yards.

2 When a question asks for or identifies "area", this means "square". For example; a bedroom that measures 12 feet by 12 feet contains 144 "square" feet. Mathematically, this rule is expressed as ft. x ft. = square feet. Remember; only units of measure can be multiplied together. Other units cannot (it's impossible because there's no meaning to it).

③ When a question asks for or identifies "volume", this means "cubic". For example; a bedroom measuring 12 feet by 12 feet with a ceiling height of 8 feet, is translated to mean the volume is 144 sq. ft. times 8 ft. or 1,152 "cubic" feet. Mathematically, the rule is expressed as ft. x ft. x ft. = cubic feet (or square feet times feet equals cubic feet).

④ Conversion of units can be confusing. For example; if the question asks how much carpeting to order for your bedroom, first determine the area or square feet. Then, the area has to be converted into yards. Remember, in this case, yard is slang for square yard. There are 9 square feet in one square yard because 3 ft. x 3 ft. = 9 sq. ft. or for specific mathematical problem solving, convert 144 sq. ft. into yards is simply done by dividing 144 by 9 = 16 yards (actually square yards). Mathematically, this is expressed as 144 sq. ft. / 9 sq. ft. per sq. yd. = 16 sq. yds.

⑤ Another situation could ask to determine how many yards of sand must be ordered and delivered to fill a children's sandbox that measures 10 feet by 12 feet by 2 feet deep. Simply multiply 10 ft. x 12 ft. x 2 ft. to determine the volume or 240 cubic feet. Then convert cubic feet into cubic yards 240 cu. Ft. / 27 cu. Ft. per cu. Yd. = 8.9 yards of sand.

6　The word "per" is an actual mathematical term. Pay attention to this term's use in a word problem. It almost always means "divided-by". So, if there's a statement that says $500 per acre, it means $500 divided by one acre. If a question comes up that asks what the price is for ¼ of an acre, it's the same as $500/1 acre x ¼ acre = $125. In essence, the term gives you a clue on how to solve the problem. For example, if a lease space costs $5 per sq. ft., the expression can also mean the total lease cost divided by the total area. Another example, concrete costs $100 per yard; thus the total cost is divided by the total yards (a type of reverse engineering).

7　Percentage can be envisioned like dollars and cents. There are 100 pennies in one dollar. So the whole amount is one dollar. The parts of the dollar are pennies. So 100% is very similar to $1.00. One percent is similar to one penny or $0.01. Ten percent is like ten pennies or a dime which is written as $0.10. 25% is like a quarter or $0.25. *Always* translate percentage into a decimal number just like what happens with money. 25 cents = 0.25 dollars and 25% = 0.25 (a simple stand alone number). The operation requires moving the decimal point over two spaces.

8   There are essential key measurement units and conversions which **<u>must</u>** be memorized.

- 43,560 square feet is equal to one acre (unit of area)

- 43,560 cubic feet is equal to one acre ft. (volume of lake water usually)

- 1 mile square (one mile by one mile is different than one square mile) = one section

- One section is equal to 640 acres or one square mile in area

- One lineal mile is equal to 5,280 lin. Ft.

- One day is 24 hours

- A statutory year is equal to 360 days (for real estate math purposes, unless noted otherwise)

- A statutory month has exactly 30 days (for real estate math purposes, unless noted otherwise)

9 *Basic* geometry formulas used in real estate mathematics are very simple to learn.

For a rectangular piece of property, the land area is equal to the front footage times the depth. The front footage is always the first measurement given. Always check the units to make sure they match and convert where necessary. For example a simple rectangular lot measuring 500 ft. by 90 yards needs converting so units match or 500 ft. by 270 ft. (500 ft. x 90 yds x 3 ft. per yd. = 135,000 sq. ft. This can easily be converted into acres. (135,000 sq. ft. divided by 43,560 sq. ft. per acre = 3.1 acres)

10 In certain cases, it becomes necessary to incorporate triangular shapes when assembling an area of a piece of land. These shapes have three sides. Generally, in basic real estate math, triangles are called right-triangles because one corner is exactly perpendicular or 90 degrees. The other two corners have any angle possible but are unnecessary in determining the triangle's area. The simple method for determining the area is basic, very similar to finding the simple area of a square or rectangle, except the triangle is half the size. So, the method is one half the length times the width. That's because a right triangle is a rectangle sliced in half going corner to corner.

Any odd shaped property can be segmented into rectangles and right triangles. Sometimes this can get complicated but for purposes of passing the licensing exam, the question normally will configure a rectangle shape with one right triangle. For example; determine the acres in a piece of property that goes 100 feet west of the point of beginning, south 100 feet, east 150 feet and back to the point of beginning. Sketch this out. It's easy to see the description makes a 100 ft. square with an adjacent right triangle measuring 100 ft. by 50 feet. Add each of these areas together. (100 ft. x 100 ft. plus ½ of 100 ft. x 50 ft. = 10,000 sq. ft. + 2,500 sq. ft. = 12,500 sq. ft.) Acreage = 12,500 sq. ft. divided by 43,560 sq. ft. per acre = 0.287 acres.

The perimeter of this piece of property is determined by "adding" the lengths of each and every property dimension. In this case, it's a little difficult to determine without understanding basic geometry (determining the hypotenuse length of the triangle). However, in real estate math, test questions stay away from this. A question may simply ask; "How much fencing is required for the 100 ft. x 100 ft. portion?" The answer is simple and can be determined by adding all the sides. 100 ft. + 100 ft. + 100 ft. + 100 ft. = 400 ft.

Exercise!
This should be fun and exciting, especially if you've never studied geometry.

You can skip this tidbit of information if you like; especially if all you are interested in is passing the state real estate exam. However, a little knowledge in the industry will go a long way.

Right Triangle is a type of code word for making things simple and understandable when learning geometry. Remember; any three sided straight-line diagram is a triangle but a right-triangle requires one of the angles to be exactly perpendicular or 90 degrees. The symbol used where the perpendicular lines occur is a small square in the corner which means 90 degrees.

So really; any non right-triangle can be divided into two parts making two right-triangles. That makes things easier when determining areas or other dimensions. One term commonly used is hypotenuse. That's always the side opposite of the right-angle (the 90 degree perpendicular corner). The other two sides are called legs.

Now it's time to disclose a remarkable phenomenon. There's something called Pythagorean Theorem which is mathematically expressed as

$$a^2+b^2=c^2$$

The letter "**c**" is referred to as the hypotenuse of a right triangle (the side opposite the right angle), sometimes referred to by students as the long side of the triangle. The other 2 sides are "**a** and **b**" (referred to as the legs of the triangle). The number "**2**" means square (like in square feet, etc.).

When there's a small number like a 2 raised up behind a figure, it means multiplying the number twice to determine area (for example). So, if **a** is equal to 5 feet, $\mathbf{a}^2$ = 5 feet times 5 feet or 25 sq. ft.

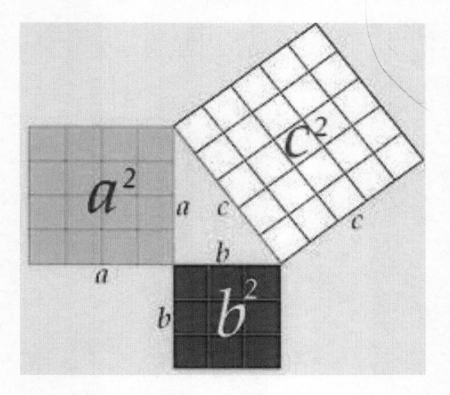

The theorem states that the square of the hypotenuse is the sum of the squares of the legs. In this image, the legs would be the sides of the triangle where **a** and **b** are. The hypotenuse is the side of the triangle where **c** is. Always understand that the Pythagorean Theorem relates the areas of squares on the sides of the right triangle.

The diagram shows the magical 3-4-5 right triangle. Carpenters use this knowledge to check corners to make sure they are square (measure out 3 feet on one wall and 4 feet on another and make sure the distance between the two locations measures exactly 5 feet).

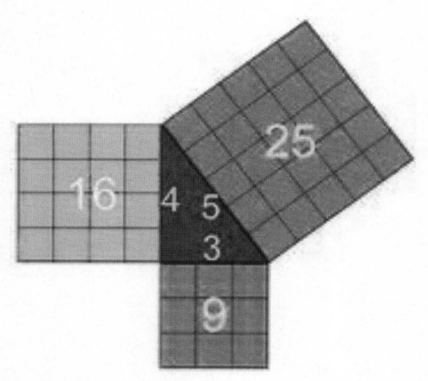

Notice in the diagram above; the squares are all equal and measure exactly the same per side (1 ft. x 1 ft. for example). So, 16 sq. ft. plus 9 sq. ft. equals 25 sq. ft. The intersection of the legs must be perpendicular (a right-triangle).

In Summary, Remember!

Converting percent into a fraction or decimal is essential. The basic rule is simply move the imaginary decimal point over two spaces (which may require adding a zero in the mix). As an example; 50% is similar to the number 50 with a decimal point at the end (50.). To convert the percent to a number, move the decimal two spaces to the left = .50. This number can be converted into a simple fraction by dividing .5 over 1.0 where both decimals can be moved one or two spaces to the right = 5/10 reduced to ½. (50% = ½)

Use the "T" formula for problems containing or asking for a percent. The basic configuration is...

A / BxC (part divided by the product of whole times the percentage)

Cover the letter that you want to determine. This will reveal the basic formula necessary to solve the equation. As an example; if "**A**" is not known – cover it with your thumb to see that **A** = **B** x **C**. If "**B**" is not known – cover it with your thumb to see that **B** = **A/C** or **A** divided by **C**. Also, "**C**" is determined similarly by covering it up; **C** = **A/B** or **A** divided by **C**. NOTE: "**A**" has a diagonal line after it which means divide (similar to a fraction that has a line - meaning the upper number is divided by the lower number). "**B**" and "**C**" are always multiplied together. All three letters represent something as follows (clues)...

A: imagine a piece of pie; say for example a small sliver cut from a whole pie. In other words, "**A**" is almost always a smaller number than "**B**". It's usually a piece of the whole thing. There's an exception, but probably only one problem on the test will require imagining a whole pie with an additional piece added, if any.

B: imagine this to be the whole pie. In other words, "**B**" is almost always a larger number than "**A**". It's the "original" whole thing before adding a piece or taking a piece away. Think of it as a freshly baked pie right out of the oven ready to start enjoying. Sometimes there's a great deal of pleasure from it and sometimes it tastes bad (like it may appreciate or depreciate, for example).

C: Always think of "**C**" as a % (for purposes of real estate math). Sometimes it can be an assessment value per $100 (which is similar to percent). It's really a number but generally has to be converted into a percentage or the percentage has to be converted into a number.

In essence, always use the magic formula to solve a word problem.

- PART/% = WHOLE
  Part divided by a percentage is equal to the whole

- PART/WHOLE = %
  Part divided by the whole is equal to the percentage

- WHOLE x % = PART
  The whole times the percentage is equal to the part

- To find the part; multiply percent by the total.

- To find percentage; divide the part by the total.

- To find the whole; divide the part by the percentage.

There are also some basic rules of thumb to employ when computing percentages.

Hint: Whenever you sense there's a word problem that describes some kind of loss (taking a piece of the pie away), it means there's a subtraction from the whole. Think of 100% as being the whole pie and if 10% of the pie is taken away, you are left with 90%. There's an easy to remember expression designed to help remind you called "**LLADD**".

L = loss, problem describes a loss in value or taking away

L = less, in other words, if some value ends up less than the original whole

A = after, the question may be asking about something after the original

D = discount, meaning the original whole is made smaller

D = deduct, meaning the whole pie has pieces deducted from it

• Sometimes a word problem will describe a percentage as being greater than 100%. If this is the case, "**A**" and "**B**" must be switched. In other words the larger number ends up on the top and the smaller number on the bottom. Keep in mind where the decimal point is placed for percentage (110% = 1.1).

- Basic (simple) questions do not need the magic formula but it still works. For example, all sales people working on commission know that if their fee is 5% of the selling price (whole pie), than simply multiply the value of the pie times 5% for the commission amount. In other words, you don't know what "**A**" is but simply multiplying **B** x **C** gives the answer.

- On more complex financial problems, more than one "T" formula is required with adding or subtracting the results to obtain the final answer.

- **Do not ASSUME.** Think what the problem is asking. For example; if a house "originally" sold for $300,000 and later sold for $400,000, the question may ask what the appreciation percentage is. Without thinking, you might immediately exercise the magic formula by placing the smaller number above the larger to get a percentage answer. Always think of the piece of pie and the whole pie which is the original. In this case the original pie is $300,000. More pieces were added to the whole because it's worth more. So find out how many more pieces of pie were added to the pie (which is a smaller number). Obviously, to determine the added pie pieces you have to subtract $300,000 from $400,000 to get $100,000. That's the number you use in the "T". However, you could plug the first values to get a percentage, but that has to be **subtracted** from the whole (100%). In both cases, the answer would be 25%.

- Residential properties may also talk about a "gross rent multiplier". This is the basic way investors compare properties. The larger the cap rate, for example, the higher the investment risk. The magic formula can also be used for this but in this case...

A = **NOI** or the net operating income (a piece of the pie). This is determined by deducting the operating expenses from the gross income to get the net income.

B = value, the whole pie (the worth of the property)

C = the cap rate expressed as a percentage (rate of return)

- Financial institutions and lenders talk about annual interest and interest rates. The "T" formula comes in handy for this.

A = the annual interest amount (the amount home owners can use for an IRS deduction)

B = the value of the loan expressed in dollar amount

C = annual interest rate

- Complex problems require using the statutory day calculations. For example, to determine the daily interest if the annual interest is $3,600; divide by 360 to get $10 per day. This is usually the first step when a problem asks for how much something is on a certain day of the year. It's important to first figure the daily amount, and then compute the number of days the problem is asking for.

Example problems with step by step explanations:

- A home owner decides to sell his home. He informs the listing agent that he has to net $87,000 or he'd have to come up with his own out of pocket expenses to close escrow.

Assuming there are closing costs of $1,000 that the seller will have and that the total sales commission will be 6% of the sale price, what will be the absolute lowest offer that the seller will accept?

SOLUTION: The first real estate math application to consider is principal item 6 using the **LLADD** reminder. There are implied words that actually mean "after". In other words, after paying the commission, the remaining amount will be $88,000 (because you also must consider the $1,000 closing costs). That being the case, the 6% sales commission must be subtracted from the whole (100%). Thus 100% less 6% is 94%. Remember, the smaller number divided by the amount or the part of the whole pie is equal to "**A**". The selling price is the larger amount, the whole pie, obviously ("**B**"). "**C**" is equal to 94% or .94. To solve the problem, the formula requires finding out what "**B**" is. Cover "**B**" in the "**T**" formula to discover that **B = A/C** = $88,000 divided by 0.94 = $93,617.

CHECK: always reverse engineer the answer to make sure it jives with what is being asked. For example, if $93,617 is correct, then take

6% of that amount to determine the actual commission (which in this case is $93,617 x 0.06 = $5,617. Then add the $1,000 closing costs, or $6,617. Subtract this amount from $93,617 and you get $87,000 net to seller. ALWAYS CHECK FINANCIAL QUESTIONS IN THIS WAY!

- A land seller must net $450,000. After paying a 10 per cent sales commission, what must he get per acre? The property description is as follows...
  Note: POB means point of beginning

  From the point of beginning at the NE corner of section 16, east 1,028 ft., south 487.34 yards, west 567.66 yards, then to the POB.

  SOLUTION: check units to make sure all match. The description is both in feet and yards. The best way is to convert all measurements to feet. 487.34 yards x 3 ft. per yd. = 1,462 feet; 567.66 yards x 3 ft. per yd. = 1,703 feet. Now, draw a sketch. It's easy to see that the configuration can be simply divided into two parcels; one is rectangular, the other a right triangle. The base of the triangle measures 1,703 feet minus 1,462feet = 241 feet. Find the area of each and add together. (1028' x 1462' plus 241' x 1462' / 2 = 1,679,107 sq. ft.). Now convert into acreage: 1,679,107 sq. ft. divided by 43,560 sq. ft. per acre = 38.55 acres.

The other component of the problem requires using the magic formula. **A** = $450,000 (since it is the net amount and the smaller number). **C** = 90% since **LLADD** applies and **B** = the selling price. Thus **B** = **A/C** = $450,000 / (90%) = $450,000 / .9 = $500,000 (what it must sell for).

The final component determines the answer to find out how much an acre this would be. Since this is simply a price per acre, the expression is the same as value divided by acreage or $500,000 divided by 38.55 acres or $12,970 per acre.

- A buyer is closing on an apartment complex July 17. The rent per unit is $450 and the July rent is already paid for all 52 units. What will the buyer be credited on closing day?

SOLUTION: remember, the closing day is the buyer's day, which translates to mean that the buyer receives the rent on that day.

Figure 30 days per month which means there are 14 days left in the month including the buyer's day (17th). Mathematically speaking, the daily rent rate can easily be determined because of the simple 2-f rule (rent is $450 per month). Thus the rent is $450/mo x 1 month/30 days = $15 per day (check: $15 per day x 30 days per month = $450 per month). Finally, $15 per day x 14 days = $210 per unit. There are 52 units, so the final answer is $210 per unit x 52 units = $10,920.

The total assessed value for all the taxable property in a city is $432,660,000. The city proposes a bond issue for city parks. The cost estimates range between 1.1 and 1.8 million dollars. What would be the average dollars of tax per $100 of assessed value?

SOLUTION: There's that word "per" again which means divided by $100 (remember this). Obviously, this will have to be a two step solution utilizing the "T" formula, because the question asks for the "average" amount. You'll need two answers and then figure the average between the two. **A** = the smaller amount and in this case, 1.1 and 1.8 million will each be plugged into the "T" formula. **B** = the larger number or $432,660,000. C is the unknown but it's very much like a percentage because its one dollar per 100 dollars so to translate "**C**" into the total dollar amount, the answer has to be multiplied by $100. For the first calculation, **C** = **A/B** = $1,100,000 divided by $432,660,000 = 0.0025 (which needs to be multiplied by $100) = $0.25 per $100 assessed value. Then do the same operation using 1.8 million (you'll get $0.42 per $100)

The average between the two answers is simply determined by adding the two and dividing the answer by 2. Thus, $0.25 + $0.42 = $0.67 and ½ of this is $0.34 (final answer).

5 MATH FOR APRAISALS

Principles and rules apply to determining a property value. The mathematics involved is crucial. The magic formula can be used in a similar way, but first some descriptions of elements and terms are essential.

An appraiser must perform field measurements and provide a diagram for documentation. Understanding legal definitions based on uniform code standards are useful in this regard.

Sometimes FOOTPRINT and FLOOR PLAN mean the same. But there's a difference in the definition.

FLOOR PLAN: A dimensioned scaled plan view of a building showing rooms and use. Various distinguishable areas are depicted such as "conditioned" and "unconditioned" spaces. For example; a garage is completely enclosed but not air conditioned or heated. A covered porch or carport is not enclosed space.

FOOTPRINT: A dimensioned plan depicting the overall ground shape, usually the perimeter foundation and covered patio slabs.

Technically, the area of a building means "livable" space. So everything enclosed by walls that are air conditioned or heated provides a "livable" area. There could be, for example, a 3 ft x 3 ft unconditioned air shaft completely inside a structure that needs to be subtracted from the gross area. Though a garage may have a washer and dryer, the space is not considered "livable". A water-heater compartment may be inside but accessed by an outside door. That area must also be subtracted if not conditioned (not livable space). Also, technically speaking, all livable or conditioned areas with a head clearance of less than five feet are not considered livable floor area. For instance, a storage area under a stair, under five feet clear, cannot be considered floor area.

The area of a building is measured from the outside. That brings up an interesting situation. Eight inch masonry walls are measured from the outside. But, a wood stud framed building with brick veneer adds an additional 5 or 6 inches to the wall thickness. That cannot be counted because it is incidental or cosmetic, and not a structural component. Floor area with conventional skeleton (stick) framing is measured to outside face of studs. That means the 1.5 inches of stucco veneer assembly should be subtracted from the wall thickness.

Most of the time, rules and legal terms for area are either not understood, or just ignored. Without uniformity, it becomes frustrating. But, for the fun of it, investigate the difference in floor area for a building 24'-3" wide by 72'-3". The field measurements included the stucco veneer but the actual framing is 24'-0" x 72'-0".

24 ft x 72 ft = 1,728 sq. ft. and 24.25 ft x 72.25 ft = 1,752 sq. ft. The difference in floor area is 1,752 sq. ft. – 1,728 sq. ft. = 24 sq. ft. When estimating value, based on $100 per sq. ft. replacement cost, the amount is $2,400. See how misleading this could become?

Commercial properties love using the term "cap rate". From the stand point of investing, the higher the cap rate the greater the risk. A comparatively lower cap rate for a property would indicate less risk associated with the investment (increasing demand for the product), and a comparatively higher cap rate for a property might indicate more risk (reduced demand for the product). Some factors considered in assessing risk include creditworthiness of a tenant, term of lease, quality and location of property and general volatility of the market.

Value is likely based on return on the money as opposed to some replacement cost figure. So the magic formula comes into play here. Understand gross income and operating expenses, common terminologies used in word problems.

If a commercial property has a gross income of $567,000 with annual operating expenses totaling $136,000, what would the property be worth if the investor is looking for a cap rate of 4.5%?

Step 1: Net income must be determined. That is essential before plugging values into the magic formula. Think in terms of investing. The risk is based on actual money earned.

Step 2: $567,000 - $136,000 = $431,000 (annual net income). This figure is less than the actual value of the property so it is a PART. The number belongs at the top (of the magic formula)

Step 3: The value (or whole) is unknown but the percentage is 4.5% which translates or converts to a number = 0.045

Step 4: Application is simply covering the unknown and discovering that the answer is $431,000 ÷ 0.045 = $9,577,777.

The higher the capitalization rate, the less the market value of the property. Capitalization rate (or "cap rate") is the ratio between the net operating income (NOI) produced by an asset and its capital cost or VALUE (the original price paid to buy the asset) or alternatively its current market value. The rate is calculated in a simple fashion as follows:

CAP RATE = NOI ÷ Value

For further clarification and understanding, let's examine a different condition. You purchase a building for $1,000,000 and it produced exactly $100,000 in positive net operating income (NOI)(the amount left over after fixed costs and variable costs are subtracted from gross lease income) during one year, then:

$100,000 / $1,000,000 = 0.10 = 10%

The asset's capitalization rate is ten percent; one-tenth of the building's cost is paid by the year's net proceeds. In other words, you made a ten percent return on your initial investment. That's a good decent return unless a less risky investment returned the same amount. These are factors investors carefully look at.

If you actually bought the building twenty years ago for $200,000, the cap rate is $100,000 / $200,000 = 0.50 = 50%.

However, you the investor must take into account the cost of keeping money tied up. By not dumping this building, you could lose the opportunity of selling and reinvesting $1,000,000. The new owner would then assume a current value, not the prior investment. That's what should be used in the cap rate calculation. Thus, even if you keep the property, which you purchased twenty years ago for $200,000, the real cap rate is ten percent, not fifty percent. Just assume you have a million dollars invested, not the original two hundred thousand.

Here's another example of why the current value should be used, consider the case of a property that is given away (as an inheritance or charitable gift). The new building owner divides his annual net income by his initial cost, say,

$100,000 (income)/ 0 (cost) = UNDEFINED

That makes no sense, obviously. From this, we see that market value, or at least an estimate, is necessary. Remember, as the value of an asset increases, the amount of income it produces should also increase (at the same rate), in order to maintain the cap rate.

Capitalization rates are, from the standpoint of an appraisal, an indirect measure of how fast an investment will pay for itself. The simplified example shows the purchased building will be fully capitalized (pay for itself) after ten years (100% divided by 10%). If the capitalization rate were 5%, the payback period would be twenty years.

Real estate appraisers use net operating

income but investors are also concerned about cash flow. Cash flow equals net operating income minus debt service. Debt service is the cost of money, such as points, interest, and handling. An investor views his money as a capital asset. As such, he expects his personal money risk to produce more money. Taking into account how much interest is available on investments in other assets, an investor generally arrives at a personal rate of return. This becomes the cap rate he expects. If an apartment building is offered to him for $100,000, and he expects to make at least 8 percent on his real estate investments, then he would multiply the $100,000 investment by 8% and determine that if the apartments will generate $8000, or more, a year, after operating expenses, then the apartment building is a viable investment to pursue.

There's also something professionals like to call direct capitalization. For example, in valuing the projected sale price of an apartment building that produces a net operating income of $30,000, if we set a projected capitalization rate at 8%, then the asset value (or price to own it) is $30,000 / .08 = $375,000. Direct capitalization is commonly used for valuing income generating property in a real estate appraisal.

One advantage of capitalization rate is that it is separate from a "market-comparables" approach to an appraisal (which compares 3 valuations: what other similar properties have sold for based on a comparison of physical, location and economic characteristics, actual replacement cost to re-build the structure in addition to the cost of the land and capitalization rates).

Given the inefficiency of real estate markets, multiple approaches are generally preferred when valuing a real estate asset. Capitalization rates for similar properties, and particularly for "pure" income properties, are usually compared to ensure that estimated revenue is being properly valued.

Capitalization rates, or cap rates, provide a tool for investors to use for roughly valuing a property based on its Net Operating Income. For example, if a real estate investment provides $160,000 a year in Net Operating Income and similar properties have sold based on 8% cap rates, the subject property can be roughly valued at $2,000,000 because $160,000 divided by 8% (0.08) equals $2,000,000.

Keep in mind, the cap rate only recognizes the cash flow a real estate investment produces and not the change in value of the property. To get the unlevered rate of return on an investment the real estate investor adds (or subtracts) the price change percentage from the cap rate. For example, a property delivering an 8% capitalization, or cap rate, that increases in value by 2% delivers a 10%

overall rate of return. The actual rate of return depends on the amount of borrowed funds, or leverage, used to purchase the asset.

As U.S. real estate sale prices have declined faster than rents due to the economic crisis, cap rates have returned to higher levels. 2010 showed 8.8% for office buildings in central business districts and 7.36% for apartment buildings.

For review; here presented is an appraisal math problem.

A commercial building has a two story main section constructed of load bearing masonry measuring 48 feet by 126 feet. Two one story wood frame wings with stucco veneer project out from the main section 26'-1 ½" to the outside stucco veneer. Each wing is exactly 24'-3" stucco face to stucco face. An elevator shaft inside, measures 12 feet x 12 feet. The market value is estimated to be $125 per square feet. The gross annual income is $346,000. What would be the net operating expenses for an expected cap rate of 7.5%?

Step 1. Determine the building area. The masonry wall section has a ground floor and upper floor area of 2 x 48 ft x 126 ft = 12,096 sf. The shaft is unconditioned and extends through both floors which need to be subtracted. 12 ft x 12 ft x 2 = 288 sf. 12,096 sf – 288 sf = 11,808 sf (for the main section). The wings are one story but there are two. The net dimension after subtracting the veneer is 24 ft x 26 ft. Floor area = 2 x 624 sf = 1248 sf. Thus the total building area is 1,248 sf + 11,808 sf = 13,056 sf.

Step 2. Determine approximate value. Based on the given price per sf estimate, the value is 13,056 sf x $125 per sf = $1,632,000.

Step 3. Convert cap rate percentage to a number. 7.5% = 0.075.

Step 4. Use magic formula. The NOI (net operating expense) is the smaller quantity which is not known. But the NOI cannot be determined without knowing net income. Thus the other two known factors are multiplied together. 0.075 x $1,632,000 = $122,400 = gross income less NOI. NOI = $346,000 - $122,400 = $223,600 (final answer).

6 RATIOS & PROPORTIONS

Proportion means a numerical relationship that compares things or people. An example of proportion is the number of girls in a class compared to the number of boys.

The popular mathematical methods of determining proportions are expressed by the following relationship.

a/b = c/d (verbally stated as A is to B as C is to D)

Comparisons are a statement of equality between to ratios. The four quantities, A, B, C, and D are said to be in proportion. Now, you might suggest that this has nothing to do with real estate math. But, you would be incorrect in this assumption.

Comparative ratios are essential in many cases. Here's a rather obscure example but explains a lot. The ratio of width to length of a table can also be the ratio of oranges to lemons in the statement that follows.

60 inches is to 36 inches as 10 oranges are to 6 lemons.

Considering arithmetic methods, a ratio is a relationship between two numbers of the same kind. In layman's terms a ratio represents how much there is of another. The numbers A and B are sometimes called terms with A being the antecedent and B being the consequent. Again, A, B, C, and D is individually referred to as terms of the proportion. A and D are called the extremes, while B and C are called the means.

The origin of the concept of ratios is unknown. Ideas probably developed because of certain things familiar to all ancient cultures. One village might have been twice as large as another so this very basic observation developed and was understood in prehistoric society.

Quantities started to be compared and that's how ratios evolved. A ratio might be physical such as speed or length, numbers of objects, or amounts of particular substances. A common example is the weight ratio of water to cement used in concrete, which is commonly stated as 1:4. This means that the weight or quantity of cement used is four times the weight of water used. The communication doesn't say anything about the total measured amounts of cement and water, nor the amount of concrete being made. Equivalently it could be said that the ratio of cement to water is 4:1, that there is 4 times as much cement as water, or that there is a quarter (1/4) as much water as cement.

A carpenter has to be familiar with ratios. A 4:12 roof pitch, as an example, doesn't mention units but the framer understands the relationship. The pitch is 4 feet above a point 12 feet back or a 4 inch rise every 12 inches (or one foot).

If there are 2 bananas and 3 pears, the ratio of bananas to pears is 2:3, and the ratio of bananas to the total number of pieces of fruit is 2:5. These ratios can also be expressed in fraction form: there's 2/3 as many bananas as pears, and 2/5 of the pieces of fruit are bananas.

Consider an orange juice concentrate diluted with water in the ratio 1:4, one part concentrate is mixed with four parts of water, giving five parts total; the amount of orange juice concentrate is 1/4 the amount of water, while the amount of orange juice concentrate is 1/5 of the total liquid. In both ratios and fractions, it is clear what's being compared. In general, when comparing the quantities of a two-quantity ratio, a fraction is derived. For example, in a ratio of 2:3, consider the amount, size volume, and number.

Ratios can be reduced (just like fractions) by dividing each quantity by the common factors of all the quantities. This method is often called "cancelling." Fractions are very much like ratios, the simplest form is considered to be that in which the numbers in the ratio are the smallest possible integers. Thus, the ratio 40:60 may be considered equivalent in meaning to the ratio 2:3 within contexts concerned only with relative quantities.

Mathematically, the figures can appear like this; 40:60 = 2:3 (dividing both quantities by 20). Grammatically, we would say, "40 is to 60 as 2 is to 3." A ratio that has integers for both quantities and that cannot be reduced any further (using integers) is said to be in simplest form or lowest terms.

***

Practical applications require some simple maneuvers. Understand cross multiplication and proportions become easy. Here is a schematic representation or formulation explaining the procedure.

$$A/B = C/D$$

Cross multiplication means **A x D = C x B**

Example: Let's say you own 40 acres of land and paid $40,000 for it. Your dear friend would like 5 acres. Based on equality only, how much should he pay you for the five acres? The answer can be determined by proportions in more than one way. You could say that 40 acres are to 5 acres as $40,000 is to an unknown amount. Thus **A** = 40 acres; **B** = 5 acres; **C** = $40,000.

One way to resolve the issue is to determine what the results are by actually dividing $A \div B = 40$ acres ÷ 5 acres = 8.0. What that means is the results of $C \div D = 8.0$. So ask yourself what amount is required for **D**? $40,000 ÷ **D** = 8. You could use trial and error until everything balances out. But cross multiplication simply requires dividing $40,000 by 8 which results in $5,000.

An alternative method would say that 40 acres is to $40,000 as 5 acres is to an unknown amount. Thus **A** = 40 acres; **B** = $40,000; **C** = 5 acres. Therefore $A \div B = 40$ acres ÷ $40,000 = 1 acre per $1,000. If that's the case, 5 acres cost 5 times the cost of one acre or $5,000.

To make computations easier, use the following methods. (may wish to memorize this)

A = **C** x **B** divided by **D**   **B** = **A** x **D** divided by **C**
C = **A** x **D** divided by **B**   **D** = **C** x **B** divided by **A**

In other words, after you correctly assemble a ratio relationship and discover what term is unknown, it's a matter of picking out the appropriate formula above to determine the answer. Use your calculator. It's that simple.

7 PRORATIONS

Corporations use this word a lot. When they tender an offer for action, the company that wants to fulfill the bid may not have enough stock or holdings to complete the deal. When that happens, a mixture of stocks and cash will be used to appease all stockholders involved. Therefore, a proration of both money and shares is granted.

Real estate lingo uses the word a little bit differently and understanding ratios can assist tremendously. Proration is used in closing a sales transaction. An independent party, or escrow, determines the balances owed by the other two parties. Proration is the amount determined for particular portions or shares of a cost or benefit to be allocated.

For example; there are two parties to a transaction, seller and buyer. If the seller has paid property taxes in advance for a full year, and the sale closes midyear, the escrow officer will give the seller a credit (amount owed by the buyer).

Real estate professionals live by rules and principles. Some of these assumptions and factors are simplified methods. The industry, as a whole, accepts the rounded procedures, though precise accuracy is legally binding in the financial arena. The following are some specific points, normally used answering questions on a state licensing examination.

☺ Buyer's Day Rule: Investor / Buyer pays for any property attachments for the closing day but receives rent from a tenant on that day.

☺ Annual Divisions: Calculations are based on a 30 day month with 12 months to a year which equates to 360 days.

☺ Property Tax: States have different rules and schedules. For the purpose of illustration, assume the first half of the total tax for the year isn't due until October 1 and the second half is due March 1.

Sample questions with explanations are the best way to illustrate pro rata.

Escrow closing occurs on March 15, 2012. The annual taxes of $1200 have not been paid. How much tax will the seller be charged at closing?

Step 1: Ratios can be applied. The simplified day factor utilizes 360 annual days. Thus the first ratio (A over B) is $1200 over 360 days. Thus A = $1200 and B = 360 days.

Step 2: The second ratio is the amount the seller owes over the number of days owing which includes January, February, and 15 days in March or 30 + 30 + 14 = 74 days. Thus D = 74 days. Remember, the day the close of escrow occurs requires the 'Buyer' to pay (not the seller). That's why March has 14 days instead of the indicated 15.

Step 3: C = A x D divided by B (using a pocket calculator). Therefore, C = $1200 x 74 days divided by 360 days = $246.67

A very precise amount can be calculated just for the fun of it to see what the difference would be using legal financial figures. There are 365 days in a year, though when including a leap year every four years, the exact number would be 365.25. Assuming 365 days, however, the number of days from the beginning of the year up to and including March 14 is; 31 + 28 + 14 = 73 days.

Step 1: The first ratio (A over B) is $1200 over 365 annual days. Thus A = $1200 and B = 365 days

Step 2: The second ratio's D = 73 days (C is unknown)

Step 3: C = A x D divided by B = $240.00 (a saving of $6.67)

Escrow closing statements show debits and credits. If the property insurance for the year is $600.00 and the seller already prepaid the year, how much will the credit be to the seller and what would the debit be for the buyer if escrow closes on February 9?

Step 1: Ratios can again be used. Since 360 days make a year in real estate language, the first ratio is $600 is to 360 days.

Step 2: Always make sure units match on the comparative ratio. If a mistake is made, and the units are upside down, the formulas will not work. Thus, the money amount at the top is unknown. The day amount at the bottom will depend on whether the answer will be a debit or credit. For example, the number of days from the beginning of the year up to and including February 9 is simply 30 days for January plus 9 days in February. But remember, the buyer is responsible for the closing day, not the seller. Reasonably then, the buyer agrees to pay for 360 days less 38 days, or 322 days worth of insurance. The seller gets credit for this amount. That being the case, the second ratio can be an unknown dollar amount over 322 days.

Step 3: A = $600, B = 360 days, D = 322 days, thus from one of the four listed formulas, C = A x D divided by B

Step 4: C = $600 x 322 days divided by 360 days = $536.67.

Step 5: Credit to the seller which is the same as debit to the buyer is $536.67.

8 LENDING

Buyers should understand something about financial markets before purchasing property. Unless the sale is a cash transaction, an encumbrance or mortgage will be recorded against the property. Besides real estate, there are other markets like bonds, commodities, derivatives, foreign exchanges, money funds, private funds, stocks, institutional, government, and retail.

Corporate financing gets complicated involving all sorts of factors like capital budgeting, financial risk management, leverage buyouts, venture capital, mergers, acquisitions, and financial planning. World capital and government consumption expenditures are other avenues of investigation concerning financial and lending practices.

Real property transactions most often involve loans which in reality is a type of debt. Like all debt instruments, a loan entails the redistribution of financial assets over time, between the lender and the borrower. The borrower initially receives or borrows an amount of money, called the principal, from the lender, and is obligated to pay back or repay an equal amount of money to the lender at a later time. Typically, the money is paid back in regular installments, or partial repayments; in an annuity, each installment is the same amount.

The loan is generally provided at a cost. In other words, the provider is asking for a profit due to the risk associated with turning funds over to another party. Fees collected are referred to as interest on the debt, which provides an incentive for the lender to engage in the risky activity.

In a legal loan, each of these obligations and restrictions is enforced by a written contract, which can also place the borrower under additional restrictions known as covenants. Monetary loans are the most common, but in practice any material object might be lent.

Acting in the capacity of a loan provider is one of the principal tasks for financial institutions. For some other institutions, issuing debt contracts such as bonds is a typical source of funding.

A secured loan is a loan in which the borrower pledges some asset (e.g. a car or property) as collateral. A mortgage loan is a very common type of debt instrument. In this arrangement, the money is used to purchase the property. The financial institution, however, is given security — a lien on the title to the house — until the mortgage is paid off in full. If the borrower defaults on the loan, the bank would have the legal right to repossess the house and sell it, to recover owed money.

There are other types of loans sometimes used in real property transactions like demand loan, subsidized loan, personal loan, or commercial loan.

***

Demand loans are short term loans that are atypical in that they do not have fixed dates for repayment and carry a floating interest rate which varies according to the prime rate. They can be "called" for repayment by the lending institution at any time. Demand loans may be unsecured or secured.

A subsidized loan is a loan on which the interest is reduced by an explicit or hidden subsidy. In the context of college loans in the United States, it refers to a loan on which no interest is accrued while a student remains enrolled in education. Otherwise, it may refer to a loan on which an artificially low rate of interest (or none at all) is charged to the borrower. An unsubsidized loan is a loan that gains interest at a market rate from the date of disbursement.

Loans can also be subcategorized according to whether the debtor is an individual person (consumer) or a business. Common personal loans include mortgage loans, car loans, home equity lines of credit, credit cards, installment loans and payday loans. The credit score of the borrower is a major component in and underwriting and interest rates (APR) of these loans. The monthly payments of personal loans can be decreased by selecting longer payment terms, but overall interest paid increases as well.

Loans to businesses are similar to the above, but also include commercial mortgages and corporate bonds. Underwriting is not based upon credit score but rather credit rating.

The most typical loan payment type is the fully amortizing payment in which each monthly rate has the same value over time requiring unequal principle pay-down amounts each month. Financial geeks use advanced mathematical methods figuring this all out. Computer software used in all escrow companies can print out an entire loan, month by month for the entire term which shows the principle and interest.

Programmed calculations are based on equations that are complex and there's no need to understand the method. Many websites have calculators that will perform the task, if required. Keep in mind that an amortization calculator is used to determine the periodic payment amount due on a loan (typically a mortgage), based on the amortization process. Repayment factors vary the amounts of both interest and principal into every installment, though the total amount of each payment is the same.

An amortization schedule calculator is often used to adjust the loan amount until the monthly payments fit comfortably into budget, and can vary the interest rate to see the difference a better rate might make in the kind of home one can afford. An amortization calculator can also reveal the exact dollar amount that goes towards interest and the exact dollar amount that goes towards principal out of each individual payment. The amortization schedule is a table delineating these figures across the duration of the loan in chronological order.

***

Predatory lending is abusive and a deceptive practice. It's a form of abuse in the financial industry. It usually involves granting a loan that puts the borrower in a position to gain advantage over him or her. Anyone involved in this particular practice could be considered a loan shark.

Usury is a different form of abuse, where the lender charges excessive interest. In different time periods and cultures the acceptable interest rate has varied, from no interest at all to unlimited interest rates. Credit card companies in some countries have been accused by consumer organizations of lending at usurious interest rates and making money out of frivolous "extra charges". Abuses can also take place in the form of the customer abusing the lender by not repaying the loan or with intentions to defraud the lender.

Real estate professionals are not attorneys and cannot give legal advice. Some states require legal consultation in a transaction. Many times, there will be questions about obligations and responsibilities concerning the loan. One impact is the income tax liability. As a general rule, the following points apply.

1. A loan is not gross income to the borrower. Since the borrower has the obligation to repay the loan, the borrower has no accession to wealth.

2. The lender may not deduct (from own gross income) the amount of the loan. The rationale here is that one asset (the cash) has been converted into a different asset (a promise of repayment). Deductions are not typically available when an outlay serves to create a new or different asset.

3. The amount paid to satisfy the loan obligation is not deductible (from own gross income) by the borrower.

4. Repayment of the loan is not gross income to the lender. In effect, the promise of repayment is converted back to cash, with no accession to wealth by the lender.

5. Interest paid to the lender is included in the lender's gross income. Interest paid represents compensation for the use of the lender's money or property and thus represents profit or an accession to wealth to the lender. Interest income can be attributed to lenders even if the lender doesn't charge a minimum amount of interest.

6. Interest paid to the lender may be deductible by the borrower. In general, interest paid in connection with the borrower's business activity is deductible, while interest paid on personal loans are not deductible. The major exception here is interest paid on a home mortgage.

Although a loan does not start out as income, it becomes income to the borrower if the borrower is discharged of indebtedness. Thus, if a debt is discharged, then the borrower essentially has received income equal to the amount of the indebtedness. The Internal Revenue Code lists "Income from Discharge of Indebtedness" in Section 61(a)(12) as a source of gross income.

Example: X owes Y $50,000. If Y discharges the indebtedness, then X no longer owes Y $50,000. For purposes of calculating income, this should be treated the same way as if Y gave X $50,000.

For a more detailed description of the "discharge of indebtedness", look at Section 108 (Cancellation of Debt (COD) Income) of the Internal Revenue Code.

***

Basic financing has a rule of thumb formula. What it says is that the principle (or the full loan amount) is the equivalent of the total interest amount times the annual interest rate x the term or length of the loan. The magic formula is $I = P \times r \times T$. P = principle; I = interest; r = rate of interest (%); T = annual time.

If the total amount of interest paid over ten years at 10% is $100,000; the principle would be P = $100,000 ÷ (0.10 x 10 years) = $100,000. In this example, the simple results indicate interest only payments so the principle is not paid off at the conclusion of the term. The total amount of principle plus interest amounts to $100,000 + $100,000 = $200,000. Over a ten year span twice the loan value is paid back.

Amortization is a better way because the principle is paid down each month. By doing this, the balance on the loan is reduced each payment cycle. This the interest rate remains the same but the amount of interest owed reduces according to the principle balance. Using a amortizing loan calculator, the equal monthly payments are

$1,321.51. In this case the total amount paid back $1,321.51 x 10 x 12 = $158,581.

For all types of loans, there's a calculator available, go to http://www.dinkytown.net/mortgage.html. Using the example above, the first month's payment includes $488.18 principal and $833.33 interest. The second month, the principal amount is $492.24. The last payment made is $1,309.95 toward the principal and only $10.92 for interest.

There are all kinds of mortgage loans. Common types are interest only, fixed rate, and adjustable rate. Other types are blended rate (where two loans with different rates are attached to the same property) mortgage and balloon (due and payable in a certain time though the loan is amortized over a long period) mortgage.

EXAMPLE PROBLEMS

- A mortgage loan in the total borrowed amount of $180,000 has a term of 30 years. If the loan is amortized monthly at the annual interest rate of 6%, what is the monthly payment?

A question like this would never show up on a state licensing exam. The reason is because professionally, it is more expedient and accurate to go on a mortgage calculator web site and plug in the values. There you will determine the answer to be $1,079.19 (final answer).

However, there's a way to get real close and possibly exact to solving a similar problem by remembering the above figures and using proportions in solving the answer to a problem with similar figures. For example...

- A mortgage loan in the total borrowed amount of $165,000 has a term of 30 years. If the loan is amortized monthly at the annual interest rate of 6%, what is the monthly payment?

  From the prior learned answer, the loan amount is $1,079.19 per month for $180,000; so using proportions, A = $1,079.19, B = $180,000, C is not known, and D = $165,000. So, from a prior chapter, C = A x D / B = $1,079.19 x $165,000 / $180,000 = **$989.26** (correct and final answer).

This is possible because the interest rate is the same. But what if the interest rate is 5%, in lieu of 6%? If that's the case, proportions can be further used. However; fluctuations are inherent through the amortization process where the ratio varies. Lowering the interest rate is not proportional so the actual answer will always be more when interest rate is lowered. Still, on a multiple choice question, an educated guess is possible based on the following criteria.

A = $989.26 and B = 6% (or 0.06)
C = unknown and D = 5% (or 0.05)
C = A x D / B = $989.26 x 0.05 / 0.06 = $824.38
(actual amount is more = $885.76)

Comment: Principle payments are combined with interest. Each consecutive month, the loan balance is decreased, thus interest payment decreases. The payment remains the same so the applied amount toward the principle is increased. So, over a 30 year or 360 total payment history, equal monthly payments toward the principle balance would simply be $165,000 / 360 = $458.33. But, the amount of interest on the full amount for the first payment is $165,000 x 0.05 / 12 = $687.50 and the applied principle amount is $198.26 which increases proportionately (360 times) until almost $850 at the end is applied to the loan amount.

Suggestion is to multiply the proportion amount by 1.07446 to get closer to the actual payment for one drop in interest below the base interest of 6%.
$824.38 x 1.07446 = $885.76 (final answer).
Note: multiply the proportion amount by 1.1117 for an additional interest rate drop (4%)

- What would the monthly payment be for a take-out mortgage in the amount of $135,000 with a 30 year term and 5% fixed interest rate?

Step 1:
A = $1,079.19 (known factor)
B = $180,000 (known factor)
C = unknown
D = $135,000

C = A x D / B = $1,079.19 x $135,000 divided by $180,000 = $809.39 (based on 6%)

Step 2:
A = 6% or 0.06
B = 5% or 0.05
C = $809.39
D = unknown

D = B x C / A = 0.05 x $809.39 divided by 0.06 = $674.49

Step 3:
1.07446 x $674.49 = $724.71 (final answer)

- What would the monthly payment be for a take-out mortgage in the amount of $123,000 with a 30 year term and 4% fixed interest rate?

Step 1:
A = $1,079.19 (known factor)
B = $180,000 (known factor)
C = unknown
D = $123,000

C = A x D / B = $1,079.19 x $123,000 divided by $180,000 = $737.45 (based on 6%)

Step 2:
A = 6% or 0.06
B = 5% or 0.05
C = $737.45
D = unknown

D = B x C / A = 0.05 x $737.45 divided by 0.06 = $614.54 (for 5% interest)
1.07446 x $614.54 = $660.30 (payment)

Step 3:
A = 5% or 0.05
B = 4% or 0.04
C = $660.30
D = unknown

D = B x C / A = 0.04 x $660.30 divided by 0.05 = $528.24 (for 4% interest)

Step 4:

1.1117 x $528.23 = $587.22 (final answer)

- George purchased a home for $167,750 and took out a mortgage for $142,375 at 8% for 30 years. What's the loan-to-value ratio on the mortgage?

Loan to value is a proportion or ratio. In other words, the loan amount compares to the mortgage amount. $142,375 is to $167,750 as 1 is to an unknown factor. Thus A = $142,375, B = $167,750, and D = 1. C = A x D divided by B = $142,375 divided by $167,750 = 0.85 or 85%. What this also means is that George put 15% down payment because 100% - 85% = 15%.

9 REVIEW

The licensing exam will have up to 10 percent of all questions related to basic mathematics. You should strive to master this subject and get all questions correct. At least that will offset any tricky questions in the other part of the test.

Here are the ten commandments of answering math related questions. Exercising this behavior will help in tackling real estate math questions.

1.   Recognize the type of problem such as commission, mortgage, simple math, or geometry.

2.   Read the entire problem carefully before even deciding on making any computations. For example, is the problem considering annual or monthly payments?

3.   Come up with a plan to solve the problem. There may be several different steps required and the results of the steps will be required to come up with the final answer.

4.   Words are confusing and some problems purposely make you think and reason. Break everything down into parts and formulas.

5.   Do not skip steps. For multiple step problems, forget doing some steps in your head. Write everything down, specifically for checking and review.

6.   Always recheck units when coming up with an answer and make sure the question is asking for the units you arrive at.

7.   Skip a question that appears too complicated and come back to it after completing the rest of the exam; that way, a fresh new look and fresh piece of scratch paper my help.

8.   Never feel all figures presented in a problem are necessary. Some questions are purposely written to confuse and make the test taker think.

9.   Guessing intelligently necessitates reasoning and eliminating answers that are not right. Remember, reverse engineering works in this regard too!

10.   Have the magic T formula memorized and write it down on scratch paper before beginning the questions.

Example problems likely to be encountered on a state examination are presented here beginning with the most basic and progressing to the more complex.

- What is the area of a parcel measuring 100 feet wide by 100 feet deep?

    A: 100 sq. ft.
    B: 0.23 acres
    C: 10,000 ft.
    D: 1,110 yd.

Solution: All answers have different units. That's the first clue the question challenges knowledge of units. Feet times feet results in square feet, so 10,000 ft. is not correct (even though 100 x 100 is 10,000). Yards can be an area when used as slang, but not ever used in the context of land area. 100 sq. ft. is a correct unit but incorrect product. Acres would be a logical choice and is the correct answer because 10,000 sq. ft. divided by 43,560 sq. ft. per acre equals 0.23 acres. B would be the correct answer.

- What is the volume of a room with a ten ft. high ceiling and measures 13.5 feet wide by 21.25 feet deep?

A: 106.26 cu. yd.
B: 2,869 sq. ft.
C: 287 cu. ft.
D: 1,350 cubits

Solution: All answers have different units. That's another clue the question challenges knowledge of units. Feet times feet results in square feet but again multiplying the height in feet results in cubic feet., so 2,869 sq. ft. is not correct (even though the numerical product is correct). Cubits is not volumetric units so that leaves A. 2,869 cu. Ft. divided by 27 cu. Ft. per cu. Yd. = 106.26 cu. Yd.

- A city's concrete reservoir measures 980 feet long by 286 feet filled three feet with water. How many acre feet of water does it contain?

A: 6.43 acre feet
B: 840,840 cu. Ft.
C: 19.3 acre feet
D: 31,142 cu. Yd.

Solution: Only two answers have the correctly depicted and asked for units. An acre ft. is a volume one foot deep, so three feet deep is 3 times as much. Thus 840,840 sq. ft. divided by 43,560 sq. ft. per acre equals 19.3 acre feet. Choose C.

- What is the total length of property-line fencing required for a parcel described as follows?

  140 feet west, then 100 feet south, then 100 feet east, then 70 feet north, then to point of beginning.

  A: 490 lineal feet
  B: 410 lineal feet
  C: 460 lineal feet
  D: 470 lineal feet

Solution: Sketch the plot out. One section is triangular and the distance unknown (along the side opposite the right angle corner or hypotenuse). Normally, test questions require no knowledge of geometry, but a simple carpenter's square is made in the field understanding that a 3-4-5 triangle creates a perfect right angle. In this example, a 30 – 40 right angle distance creates a 50 ft. long hypotenuse. Thus, the perimeter is the sum of 50 feet + 140 feet + 100 feet + 100 feet + 70 feet

= 460 feet. Choose C (correct answer).

- What is the area of land with the following measurements?

  178 feet west, then 934 feet south, then 78 feet east, then 134 feet north, then to point of beginning.

  A: 26 acres
  B: 2.6 acres
  C: 166,250 ft.
  D: 112,852 ft.

Solution: Answers have different units. That's the first clue the question challenges knowledge of units. Feet times feet results in square feet, so two answers fail to satisfy this requirement except acres measures land area as well. Sketch the plot to again see there are two basic shapes comprising a right triangle and rectangle. The triangle measures 100 feet by 800 feet = 80,000 sq. ft. divided 2 = 40,000 sq. ft. The rectangular section measures 78 feet by 934 feet = 72,852 sq. ft. Combined, the property area = 40,000 sq. ft. + 72,852 sq. ft. = 112,852 sq. ft. Do not choose D, because the units show ft. and not square feet. Choose B as the correct answer because converting to acres is easy, just divide the sq. ft. by 43, 560 sq. ft. per acre.

- What is the area of a parcel measuring 100 feet wide by 100 yards deep?

    A: 10,000 sq. yd.
    B: 0.69 acres
    C: 30,000 ft.
    D: 90,000 sq. ft.

Solution: All answers have different units. That's the first clue the question challenges knowledge of units. Feet times feet results in square feet, so yards must be converted to lineal feet. 3 feet per yard times 100 yards is 300 feet. 300 ft. x 100 ft. = 30,000 sq. ft. (so C is incorrect); but converting to acres results in 0.69 acres.

- How many yards of concrete is required when it is best to order 10 percent more than the actual slab measuring 20 feet by 24 feet by 4 inches thick?

    A: 6.0 cu. yd.
    B: 6.65 yards
    C: 160 yards
    D: 66.5 yards

Solution: Cubic yards of concrete is typically referred to as yards in the industry. The volume of concrete slab can be determined by converting units to match (4 inches = 0.34 ft.). 20 ft x 24 ft x 0.34 ft = 163 cu. ft. which is converted into cubic yards by dividing by 27 cu. ft. per cu. yd. = 6.0 cu. yd. But 'A' would not be the correct answer since the suggestion is to add 10% to the order. 0.10 x 6 cu. ft. = 0.6 cu yds. Total is 6.6 yards. Choose B.

- Order carpet for a house figuring a 15% overage for waste. The total area to be covered is 1,200 sq. ft. and the price is $16.78 per yard. What is the total cost?

A: $2,208.56
B: $2685.00
C: $2237.34
D: $2573.00

Solution: 15% = 0.15 and the total overage to order would be 1.15 x 1,200 sq. ft. = 1,380 sq. ft. Convert this to square yards = 1,380 sq. ft. / 9 sq. ft. per yard = 153.34 sq. yd. Multiply the cost per yard times the total ordered yardage = $16.78 x 153.34 sq. yd. = $2573.00. Choose D.

- The sale commission is 10% for a property that sells for $1,050 per acre of land that measures exactly ¼ mile x 1,940 yards. How much commission did the listing agent receive if the selling agent made 60% of the commission?

A: $74,072.00
B: $11,851.60
C: $118,516.09
D: $7,407.20

Solution: Converting units to match is critical and since there's 43,560 sq. ft. to an acre, converting all dimensions to feet is recommended. ¼ mile is the number of feet in a mile divided by 4 = 5,280 ft. / 4 = 1,320 ft. Convert yards to feet by multiplying by 3 ft. per yd. = 1,940 yd. x 3 ft. per yd. = 5,820 ft. The area of land in sq. ft. = 5,820 ft. x 1,320 ft. = 7,682,400 sq. ft. Convert to acres = 7,682,400 divided by 43,560 sq. ft. per acre = 176.364 acres.
Total sales price = 176.364 x $1050 = $185,182.20. The total commission is 10% = $18,518 and 40% is equal to $7,407.20. Choose D.

FYI: Engineers prefer composing a complete formula in this endeavor. The composition would result in the identical answer but the operation would be one step using a scientific calculator.

$$\frac{0.25 \times 5{,}280 \text{ ft/m} \times 1{,}940 \text{ yd.} \times 3 \text{ ft./yd.} \times \$1{,}050/\text{acre}}{43{,}560 \text{ sq. ft. /acre}} \times 0.10 \times 0.4 = \$7{,}407.20$$

- Jane borrows $45,000 at 6% simple interest for 3 years. What is the total principle and interest paid at the conclusion of the loan.

    A: $49,284
    B: $53,100
    C: $47,700
    D: $52,009

Solution: The clue to solving the problem is the definition of "simple-interest" which means just determining the annual interest and multiplying that amount by 3 years. Compound interest is different and would entail amortizing over the three year period. Monthly payments, in that case, would be $1369 and over 36 months would be $49,284. But the correct answer assumes paying the borrowed money's interest in total and annually, the interest only payment equates to 6% times $45,000 x 3 years = $8,100. Add that amount to the payback principle to arrive at the answer of $53,100. Choose B.

- If a property is purchased for $550,000 and later resold for $672,000. What percentage of profit was earned?

A: 100 sq. ft.
B: 22.18%
C: 10,000 ft.
D: 1,110 yds.

Solution: In this situation, there are a two ways to solve the problem.

1. Using the magical or "T" formula, the unknown is the percentage (%) and the smaller number, in this case, doesn't belong at the top like most of the time because the answer, in essence is larger than 100%. That's because the *after* sold price is more than the original. The answer, using this option, is $672,000 divided by $550,000 or 1.2218. The percentage of profit is 22.18% greater than the whole. The acronym (LLADD), where A = after, is the clue that the answer is subtracted from 100%.

2. The other way, using the magical T, means percent is equal to the part divided by the whole. The part is the difference between the two selling prices and the whole is the original or base *whole* amount. That being the case, the increased amount becomes the part and thus % = $122,000 divided by $550,000 = 0.2218 = 22.18%.

- An unimproved subdivision lot measures 100 feet by 200 feet. Value in the area is based on the front footage. If the property sells for $456,000, how much per foot is the front footage worth?

    A: $22.80
    B: $4,560.00
    C: $2,280.00
    D: $280.00

Solution: Front footage is always the first dimensional figure stated. The second figure is not even relevant, to answer the question. Always understand specifically what he question is and never jump to conclusions, such as figuring the lot area. The answer simply understands the price per foot is the price divided by feet. Thus, the answer is determined by taking the full price and dividing by the front footage length = $456,000 divided by 100 feet = $4,560.00 (per lineal foot). Choose B.

- A buyer purchased a home for $166,000 exactly five years ago. The property depreciated 2.5% each year. That being the case, what is the current value of the home?

    A: $146,262
    B: $145,250
    C: $186,750
    D: $191,089

Solution: Depreciation means the value is less, not more; so the answer must be less than the original purchase. Two different means to the answer are possible. One is precise, the other is approximate. Generally, state exams are simple means to an end. If that's the case, the first method applies.

1. Multiply 2.5% times the five years and then subtract the results from the base price for the overall depreciation. $166,000 x .025 x 5 = $20,750. Subtract the depreciation from the base price. $166,000 - $20,750 = $145,250. Choose B

2. The precise method is cumbersome because after each year, the value of the property decreases 2.5%. That essentially means the value the following year decreases by a less amount because the 2.5% is based on the second year value and not the original. The configuration is as follows...

   $166,000   decreases to $161,850 (year 1)
   $161,850   decreases to $157,804 (year 2)
   $157,804   decreases to $153,859 (year 3)
   $153,859   decreases to $150,012 (year 4)
   $150,012   decreases to $146,262 (year 5)
   The precise answer is A.

- A house originally sells for $236,000. The next time it sells, the price is $321,000. What was the percentage of appreciation?

A: 36%
B: 73.5%
C: 26.5%
D: 64%

Solution: Appreciation is of course the term used for increase in value. There are two ways to solve the problem.

1. Determine the part which is the difference between the current and original sales price. $321,000 - $236,000 = $85,000. The whole is the <u>original</u> base price. Thus, the percentage is determined by taking the part and dividing by the whole = $85,000 / $236,000 = 0.36 = 36%.

2. The same answer is determined by understanding a percentage can be larger than a whole (more than 100%). That being the case, use the larger amount and divide it by the smaller amount since the magical "T" formula's part in this case is larger than the "original" whole price. $321,000 divided by

$236,000 = 1.36 or 136% but subtract 100% to understand the increased appreciation, which is 36%. Choose A.

- A property is worth $234,000 and has a county assessment ratio of 11.5%. What would be the total annual property tax if the rate is set at $1.66 per $100 of assessed value?

A:	$4,467
B:	$1,621
C:	$8,158
D:	$446.71

Solution: A beginning understanding of the question requires understanding tax "units" which is basically the assessed or actual appraised market value divided by the $100 factor. But don't forget there's a tax rate to consider as well.

1. First figure the tax units which would be $234,000 divided by $100 = 2,340 tax units. The assessment ratio must be translated from percent to a decimal = 0.115. The tax is $1.66 x 0.115 x 2,340 = $446.71.

2. Another way is correctly interpreting the word "per" in the question. Whenever this

shows up, it means divide. So the tax rate is correctly interpreted to mean $1.66 divided by $100 which is 0.0166. But the assessment ratio is 11.5% or 0.115. Thus the annual tax is based on a value of $234,000 x 0.115 = $26,910. The tax is $26,910 x 0.0166 = $446.71.

3. A one step mathematical procedure is figured as follows...

$234,000 x 0.115 x $1.66/$100 = $446.71. Choose D.

- A buyer and seller agree to a closing date of February 9. The annual insurance premium amounts to $669 and has been prepaid for the year. How much does the buyer owe the seller?

  A: $596.50
  B: $598.38
  C: $591.18
  D: $70.70

Solution: Two ways are presented, the reasoning method and the exact way.

1. The daily insurance cost is figured to be a statutory 360 day year. That being the case,

the cost per day is reasonably determined because "per" means divide by. In this case, $669 divided by 360 days = $1.86 each day. The number of days the seller owes for is 30 days for January (since a statutory month has 30 days) plus 9-1 days or 8 days (since the "buyer" always owes for the day of closing). The total owed by the seller is 38 days times $1.86 = $70.68 which is the debt to seller. It thus stands to reason, the amount the buyer owes would be the balance between the annual amount less the seller's contribution = $669 - $70.68 = $598.32

2. The exact way is a one step formulation as follows...
$669 - ($669/360 days) x (30+9-1) = $598.38   Choose B

• A business enterprise owns property that has a gross income of $346,000 with $98,500 in operating expenses. Assuming a cap rate of 6%, what is the property worth?

A: $4,125,000
B: $412,500
C: $3,723,450
D: $262,350

Solution: The magical "T" formula can be utilized

where the cap rate (rate of return) is positioned where percent is. The value is the whole and the NOI (net operating income) is the part. Two methods are discussed here.

1. NOI is determined by deducting expenses from the gross income. $346,000 -$98,500 = $247,500 = NOI. Using the magic "T", NOI divided by the % (cap rate) determines the value. $247,500 divided by .06 = $4,125,000.

2. The one step method is as follows...
($346,000-$98,500) / 0.06 = $4,125,000.
Choose A

- A house is valued at $191,000 and the lot is worth $29,000. What was the value of the house exactly 7 years ago when brand new if it depreciated 5% per year?

A: $257,850.00
B: $338,461.54
C: $273,506.80
D: $293,846.15

Solution: Appreciation is of course the term used for increase in value. Depreciation means less (not more). There are two ways to solve the problem (exact and approximate). Using the approximate method is advisable for test purposes to see if the answer is listed. If another answer is less, the correct choice may require figuring out utilizing the exact procedure outlined below.

1. 7 years ago means depreciation of 5% x 7 years = 35% total. The value of the land means nothing and should not be considered. The whole (the "original" amount) is unknown. The part is $191,000 and the percentage is actually 100% less 35% or 65%. Discussing percentages, reason out what the question is asking. 35% depreciation means 65% drop in value. So using the magical "T" formula, the part is divided by the percent to determine the whole. $191,000 divided by 0.65 = $293,846.15.

2. It's always a good idea to check answers by reverse engineering. In other words, the home is worth $293,846.15 and what would it be worth is it depreciated 5% every year for 7 years? The simple solution requires multiplying 7 years times 5% and then multiply that answer by the value of $293,846.15 = $102,846.15. Subtract from the $293,846.15 = $191,000. Note: the reverse process shows how "subtraction" is

necessary which the original step required when deducting the 35% from 100%.

3. Exact method, in this case, is unarguably complex, especially when the statement specifically states 5% per year and not 35% over 7 years. Working out the details means developing a 7 cycle table.

(7) $191,000   / 0.95 = $201,052.63
(6) $201,052.63 / 0.95 = $211,634.35
(5) $211,634.35 / 0.95 = $222,773.00
(4) $222,773.00 / 0.95 = $234,497.89
(3) $234,497.89 / 0.95 = $246,839.89
(2) $246,839.89 / 0.95 = $259,831.46
(1) $259,831.46/ 0.95 = $273,506.80

Always select a multiple choice answer which is closest to the exact answer.
Choose C

- The lender for a purchased home for $289,900, approves the new owner for a 90 percent loan. The earnest money deposit amounts to $15,000. How much more is necessary to complete the transaction?

A: $24,741   C: $4,399
B: $13,990   D: $15,000

Solution: A three step process is required to solve the given problem. Keep in mind that the unknown amount is the "part".

1. The part is equal to the whole times the percentage. Thus, $289,900 x 90% = $289,900 x 0.90 = $260,910.

2. The difference between the value and the loan amount is $289,900 - $260,910 = $28,990.

3. The difference between the down-payment and the earnest money deposit is $28,990 - $15,000 = $13,990.

   A one step process using an engineering calculator looks like this.
   $289,900 – ($289,900 x 0.90) - $15,000 = $13,990. Or:
   $289,900 x (1.00 - .90) - $15,000 = $13,990
   Choose B

- A buyer agrees to pay $175,000 for a home. A one point loan origination fee is required and there's also the necessity to pay 2.5 discount points to receive the bargain basement interest rate. If the buyer receives a 90% loan-to-value ratio, how much will the buyer owe at closing for points?

A: $5,512.50
B: $23,625.00
C: $6,125.00
D: $3,987.25

Solution: Complex wording can become confusing, but all that is asked is how many total points there are and what it amounts to. In other words, the 90 percent factor is necessary, because the solution to the problem requires computing the actual loan value; points are paid on the loan, not on the value!

1. 1 point plus 2.5 points is 3.5 points total. A point is the same as a percentage. So convert 3.5% = 0.035.
2. The 90% ratio means 90% times the value = 0.90 x $175,000 = $157,500.

3. The total amount of points as a percentage of the loan amount = $157,500 x 0.035 = $5,512.50.

   One step operation with an engineering calculator =
   ($175,000 x .90) x (0.025+0.01) = $5,512.50
   Choose A

- A net listing simply means the seller absolutely requires a certain amount from the sale. In one case, the seller must receive $79,900 to pay off the mortgage. The agent wants 6% commission. What would the minimum sales price have to be?

A: $88,988
B: $84,694
C: $85,000
D: $92,387

Solution: (Hint; LLADD). Always subtract percentage from 100% when the question refers to something "after": in other words, the amount left after a commission is paid.

1. 100% - 6% = 1.00 – 0.06 = 0.94

2. The whole is the amount the home must sell for and unknown. The part is the net amount. Thus the whole is equal to the part divided by the percentage. $79,900 divided by 0.94 is equal to $85,000.

3. Reverse engineer, in other words, check the answer to make sure it works. $85,000 x 6% sales commission = $5,100. Subtract the commission from the sales price = $85,000 - $5,100 = $79,900 (net price for net listing). Choose C

- Joe, a real estate agent, sold a property for $150,000. He was paid 7% commission on the first $50,000 and 5% on everything over that amount. How much was he paid?

  A: $8,500
  B: $18,000
  C: $7,500
  D: $10,000

Solution: Convert percentages into numbers and then go through the steps.

1. 7% of $50,000 = 0.07 x $50,000 = $3,500.

2. $150,000 - $50,000 = $100,000 (remaining)

3. 5% of $100,000 = $5,000

4. Total commission = $3,500 + $5,000 = $8,500.

   Choose A

- A commission check was $3,280. The sales agent received 7% of the first $30,000 and 5% on the remaining balance. What was the total sales price?

A: $23,600
B: $53,600
C: $43,890
D: $61,235

Solution: One way to figure out the correct answer is to apply reverse engineering on the given answers to see which one passes the scrutiny. Here presented are the correct steps.

1. Determine the total commission amount on 7% of $30,000. 0.07 x $30,000 = $2,100.

2. The difference between the full commission check and the fixed amount is equal to $3,280 - $2,100 = $1,180.

3. The whole is unknown. The part is $1,180 and the percentage is 5%. Thus the whole is equal to the part divided by the percent = $1,180 divided by 0.05 = $23,600.

4. The sales price is the total amount or $23,600 + $30,000 = $53,600.

   Reverse engineer (to check) means subtracting the fixed amount of $30,000 from the total sales price = $23,600 and multiplying by the 5% commission = $1,180. Add this to the fixed commission = $1,180 + $2,100 = $3,280 (commission check) Okay! Choose B

- Agent Joe received $937.50 as a commission for a referral fee. The listing agent paid 25% of the commission for the referral. The total commission paid was 6%, which was split equally between the listing and selling brokerage firms. If the listing agent was paid 50% of her firm's commission, what was the sales price of the property?

A: $150,000
B: $250,000
C: $125,000
D: $350,000

Solution: Another complexity involving several steps means carefully understanding the question and circumstances. Each step represents a segment of the final computation. What is crucial involving a wordy problem is carefully looking at the particular stipulations. For instance, the listing agent was paid 50% or half the 6% commission.

1. Determine the total commission amount paid. If Joe's commission is part of the total representing the percentage of 25% of the whole, than the part is divided by the percentage to determine the whole amount.

Thus, $937.50 divided by 0.25 = $3,750. Check this reasoning by reversing the procedure. $3,750 x 25% = $937.50 (okay).

2. Since $3,750 represents 50% or half of the 6% of the sales price, logically, the total commission is twice the amount of $3,750. That being the case, the total commission amount representing the 6% commission is $3,750 divided by 0.50 = $7500.

3. The percentage is known and the part is $7500. Thus, the whole = part divided by the percent = $7500 divided by 0.06 = $125,000. Note: a mistake could occur here by dividing the $7500 total commission by 3% which is not the **total** commission paid.

    Check the answer by reversing the conditions. $125,000 x 6% commission is $7,500. Half of this commission goes to the listing agent and the other half goes to the selling agent. $7,500 / 2 = $3,750. The referral fee is based on 25% of this amount = $3,750 x 0.25 = $937.50 (okay).

- The S 1/2 of the NE ¼ and the NW ¼ of the SE ¼ contains how many square feet?

A: 2,344,560
B: 3,484,800
C: 4,395,340
D: 5,277,200

Solution: Understand there are 640 acres in a section. The actual acreage in these two described sections are as follows.

1. 640 acres divided by 4 and again divided by 2. Or multiply the denominators described in the legal (4 x 2 = 8) and use this figure to divide by. Thus 640 acres divided by 8 = 80 acres.

2. The second description is 4 x 4 = 16 for the denominator or 640 acres divided by 16 = 40 acres.

3. The total area = 40 acres plus 80 acres = 120 acres.

4. Convert acres into square feet to determine answer. There is 43,560 sq. ft. per acre. Proportions can be utilized. One acre is to 43,560 sf as 80 acres are to ? Thus 80 acres = 43,560 sf x 120 acres = 5,227,200 sq. ft. Choose D

- George paid 20 percent down payment and secured a 30 year loan at 7% interest rate. The first payment made included $787.50 in interest. How much did he pay for the property?

A: $135,000
B: $140,625
C: $168,750
D: $158,250

Solution: Amortized loans start out with the full value on the first interest payment. Thus, the answer is easily determined using the following steps.

1. $787.50 is equal to 1/12 of the annual interest. Thus, the total interest paid on the full amount is equivalent to 12 x $787.50 = $9,450.

2. Percent is known and the part is known. Thus the whole is equal to the part divided by the percent. $9,450 divided by 7% = $9,450 / 0.07 = $135,000 (loan amount).

3. 20 % down means the $135,000 loan is equivalent to (100% - 20%) 80% of the whole. Thus the whole = the part divided by

the percent = $135,000 divided by 0.80 = $168,750.

Choose C

- A home buyer negotiates a loan for $125,000 amortized for a twenty year term at 7% fixed interest rate. Monthly principal and interest amounts to $969.12. How much interest will be paid over the life of the loan?

A: $133,750.09
B: $175,000.00
C: $107,588.80
D: $232,588.80

Solution: There's no trick or anything complicated about this question.

1. 20 years requires 12 x 20 monthly payments over the life of the loan = 240 payments.

2. $969.12 x 240 payments = $232,588.80 (includes all interest and payments on the principal amount).

3. Interest is determined by simply subtracting the total principal (original loan amount). $232,588.80 - $125,000 = $107,588.80. Choose C

Gerald Shingleton is an author, architect, and real estate developer. Semi retired in architecture leaves lots of moments for writing. He is a graduate of the Architecture School at California State Polytechnic University in San Luis Obispo, in Central California. Volunteerism is a major role in his life.

51114000R10080

Made in the USA
Lexington, KY
12 April 2016